William Dawson

Modern Ideas of Evolution as Related to Revelation and Science

William Dawson

Modern Ideas of Evolution as Related to Revelation and Science

ISBN/EAN: 9783744653701

Printed in Europe, USA, Canada, Australia, Japan

Cover: Foto ©berggeist007 / pixelio.de

More available books at **www.hansebooks.com**

MODERN IDEAS

OF

EVOLUTION

AS RELATED TO

REVELATION AND SCIENCE

BY

SIR J. WILLIAM DAWSON, C.M.G. LL.D. F.R.S. &c.

AUTHOR OF

ACADIAN GEOLOGY' 'THE CHAIN OF LIFE IN GEOLOGICAL TIME
'EGYPT AND SYRIA, THEIR PHYSICAL FEATURES IN
RELATION TO BIBLE HISTORY' ETC.

FOURTH EDITION

LONDON

THE RELIGIOUS TRACT SOCIETY

56 PATERNOSTER ROW, 65 ST. PAUL'S CHURCHYARD
AND 164 PICCADILLY

PREFACE

THE object of this work is to examine in a popular manner, and to test by scientific facts and principles, the validity of that multiform and brilliant philosophy of the universe which has taken so deep hold of the science and literature of our time. The task is a somewhat ungracious one, especially in England, whose people are naturally proud of discoveries and generalisations which, originating among themselves, have taken the world by storm. It is also extremely difficult, because of the dazzling and attractive nature of the hypothesis of evolution, the dashing and plausible character of the arguments by which it is sustained, and its all-embracing scope, which enables it to account for everything that has previously been mysterious. Besides this, it is of the nature of this protean philosophy that it should itself be in process of evolution from day to day, and thus to be in so rapid motion that it changes its features momentarily while one endeavours to sketch it.

Why then attempt such a task? The answer is two-fold—general and personal. First, the world of general readers is captivated, dazzled and perplexed by the new philosophy, and greatly needs some clear and intelligible exposition of its nature and tendency, some classification of its variations, and some attempt to explain its agreement or discordance with science and religion. Secondly, the writer of the following pages has of late years been besieged by so many letters and inquiries respecting this subject, to which he has incidentally referred in popular books on science, that it becomes necessary in self-defence and to save time to prepare an answer which may meet all demands of this kind.

The conclusions which he has reached as the result of much reading and reflection, as well as of a long-continued and somewhat wide and varied study of nature, may not satisfy the present excitement of enthusiastic specialists and lovers of novelty, but they may serve somewhat to mitigate present extremes of feeling and belief, and may accord with the sober second thoughts which sometimes follow sudden revolutions.

J. W. D.

1890.

CONTENTS

MODERN IDEAS OF EVOLUTION

CHAPTER I

PRESENT ASPECTS OF THE QUESTION

THE great fabric of the Darwinian evolution may be said to have attained to its completion. Its chief corner-stone has been laid with shouting by its jubilant adherents, and it is presented to us as a permanent and finished structure, fitted to withstand all the attacks of time and chance. We are even asked to regard its architect as the Newton of Natural Science, and to believe in the finality and completeness of the structure which he has raised.

In seeming contrast with this, we find that the disciples of the great teacher are already beginning to diverge widely in their beliefs, and to found new schools, some of which are tending toward the old and discarded theory of Lamarck, or to a modification of it known as Neo-Lamarckianism, while others boast that they maintain the pure Darwinian doctrine, though even among these there are diverse shades of belief. Thus, like other hypotheses and philosophical systems which have preceded it, Darwinism seems to have entered on a process of disintegration, and it is not easy to divine in what form or forms it may be handed down to our successors.

While thus liable to different interpretations within itself, the Darwinian evolution has still more varied aspects when we regard it in relation to the other beliefs and interests of humanity. The hypothesis has been applied to all sorts of uses in relation to physical and natural science, as well as to history and sociology, and it has been made a means of revolutionising our classifications and our ideas of species and other groups. It is sometimes monistic or positivist, and scarcely distinguishable from the old-fashioned atheism and materialism. Sometimes it assumes the newer form of agnosticism, and poses as neutral and indifferent with regard to those spiritual interests of man which are important beyond all others. Again, it becomes theistic, and here we have adherents of the new system ranging from those who are content to reconcile it with a theistic belief, which recognises a God very far off and shorn of His more important attributes, to those who accept evolution as a new gospel, adding fresh light to that which shines in the teaching of Jesus Christ. At a lower level it is evident that the ideas of struggle for existence and survival of the fittest, introduced by the new philosophy, and its resolution of man himself into a mere spontaneous improvement of brute ancestors, have stimulated to an intense degree that popular unrest, so natural to an age discontented with its lot, because it has learned what it might do and have, without being able to realise its expectations, and which threatens to overthrow the whole fabric of society as at present constituted.

In these circumstances it seems desirable that science, and especially natural and physical science, which may in some degree be held responsible for this movement, should define its own position, and do what it can to remove the difficulties and relieve the fears which have been engendered by the use or misuse of its facts and principles.

Science will in this way best consult its true interests ; since, if it commits itself to a philosophy professing finality, it is pretty certain to suffer in the inevitable reaction. On the other hand, if it will carefully sift that which is true from that which is false or hypothetical, it may ultimately fall heir to anything that may be valuable or permanent in the new philosophy without suffering from its mistakes.

We must bear in mind in this connection, that systems of philosophy which endeavour to explain everything by one idea, as they have appeared from time to time, though they have sprung into the field like boastful Goliaths, cowing too many good men for a time into silence or retreat, have soon proved vulnerable to mere pebbles from the armoury of nature.

Those especially whose studies of philosophy began half a century ago, and who have seen several such systems wax and wane, besides knowing that the same process has been going on ever since the time of Thales of Miletus, have lost confidence in the infallibility of such all-embracing generalisations, and may be pardoned for at least cautioning their younger colleagues against sacrificing science to speculation, and against the tendency to become merely scientific specialists without breadth or sympathy for higher things.

The example of the great apostle of evolution himself should warn us as to this. Darwin, as he sits in marble on the staircase of the British Museum, represents a noble figure, made in the image of God, and capable of grasping mentally the heaven above as well as the earth beneath. As he appears in his recent biography, we see the same man paralysed by a spiritual atrophy, blinded and shut up in prison and chained to the mill of a materialistic philosophy where, like a captive Samson, he is doomed to grind all that is fair and beautiful in nature into a dry and formless dust. Would that he had lived to pull down the temple of Dagon with his own hands, even if an ephemeral reputation had perished in the ruins, and to avenge himself of the cruel enemies that had put out the eyes of his higher nature!

This depth of unscientific and unspiritual degeneration, into which the mind may be thrown by the excessive pursuit of evolutionary ideas, is well shown

by Darwin himself in a letter written a year before his
death. With reference to his doubts as to the existence of God, he asks—'Can one trust to the convic-
tions of a monkey's mind ?' But if the idea of God
may be a phantom of an ape-like brain, can we trust
to reason or conscience in any other matter? May
not science and philosophy themselves be similar
fantasies, evolved by mere chance and unreason? In
any case, does not this deprive science of the ennobling
idea that nature is the development of Divine Mind,
and so reduce it to mere drudgery, pursued only for
its useful applications or for self-interest?

This seems a serious indictment against evolution,
at least in its extreme forms, but its validity seems to
be proved by a careful scrutiny of the developments
that have followed the publication of the *Origin of
Species*, and which, despite the efforts of so-called
theistic and Christian evolutionists, may be held to
have tended constantly to a lower and lower depth of
materialistic agnosticism, and, at the same time, of
debasement of natural science into a jumble of false
classifications and visionary speculations. Neither
science nor theology need, however, slide hopelessly
into this gulf, and it may even be possible to stand
near to the treacherous margin and to rescue some
grains of truth from this 'confused movement of the
mind of our age,' as it has been called by a recent
German writer.[1]

[1] Wiegand, *Darwinismus*, notice in the *Academy*, Aug. 25, 1877.

In endeavouring to secure this desirable result, we must not take for granted the truth of the assertion so often confidently made, that science is hostile to religion. It is no doubt true that monistic and agnostic evolution, and those forms of Darwinism which follow the author of the system in negation of the living God, are inconsistent with religion as well as with all the higher interests of men. There may, however, be a theistic principle of development apparent in all nature, and which represents what we can perceive of the plan and methods of creation, understanding by that word the making of all things by Almighty Power, whether immediately or mediately, through means of things already made, and laws previously established. It may be said in favour of this view that it gives an inexpressible dignity to man and to science. It shows that the human reason must be after the model of the infinite Divine reason, that in scientific inquiry we are studying God's laws and revelation of Himself in nature. Nay, more, if we regard Christ as an incarnation of the Creator, we have in Christianity itself a higher revelation of God, which must be in harmony with nature ; and we shall have a right to hold that the scientific investigator is doing Christ's work and God's work, and, on the other hand, that those qualities of humility, faith, sincerity, and love of truth which God requires of His followers are also those most profitable in scientific study, while scientific habits of thought are of the utmost value in the study of revelation and in the difficulties of the Christian life.

It is also to be observed that even the positivist and agnostic admit, as appears in recent controversies, that some religion or substitute for it is necessary to the highest perfection of man. For example, Harrison, in a recent paper,[1] believes as a positivist in what he calls the religion of humanity, that is, in setting up an ideal standard of human nature, based on historical examples, as something to live up to. His opponent Huxley, from the point of view of an agnostic, thinks this futile—stigmatises man as a failure, and as a ' wilderness of apes '—and would adore the universe in all its majesty and grandeur.

In this they rehabilitate very old forms of religion, for it is evident that the most ancient idolatries consisted in lifting up men's hearts to the sun and moon and stars, and in worshipping patriarchs and heroes. Thus we find that there can be no form of infidelity without some substitute for God, and this necessarily less high and perfect than the Creator Himself, while destitute of His fatherly attributes. Further, our agnostic and positivist friends even admit their need of a saviour, since they hold that there must be some elevating influence to raise us from our present evils and failures. Lastly, when we find the ablest advocates of such philosophy differing hopelessly among themselves, we may well see in this an evidence of the

[1] *Nineteenth Century.*

need of a divine revelation. Now, all this is precisely what the Bible has given us in a better way. If we look up with adoring wonder to the material universe, the Bible leads us to see in this the power and Godhead of the Creator, and the Creator as the living God, our Heavenly Father. If we seek for an ideal humanity to worship, the Bible points us to Jesus Christ, the perfect man, and at the same time the manifestation of God, the Good Shepherd giving His life for the sheep, God manifest in the flesh and bringing life and immortality to light. Thus the Bible gives us all that these modern ideas desiderate, and infinitely more. Nor should we think little of the older part of revelation, for it shows the historical development of God's plan, and is eminently valuable for its testimony to the unity of nature and of God. It is in religion what the older formations are in geology. Their conditions and their life may have been replaced by newer conditions and living beings, but they form the stable base of the later formations, which not only rest upon them, but which without them would be incomplete and unintelligible.

The lesson of these facts is to hold to the old faith, to fear no discussion, and to stand fast for this world and the future on the grand declaration of Jesus—'God so loved the world that He gave His only begotten Son, that whosoever believeth in Him should not perish, but have everlasting life.'

It is somewhat reassuring that the controversies

respecting evolution centre around the Bible, which is thus shown to be a formidable power in the world, and not a thing of the past, as some would have us suppose. [In this connection it is to be observed that the attitude of the Bible is often misrepresented, since, though it affirms distinctly the creation of all things by the living God, it does not commit itself either as to the limits of species or as to any special doctrine with respect to the precise way in which it pleased God to make them. When we look at the details of the narrative of creation, we are struck with the manner in which the Bible includes, in a few simple words, all the leading causes and conditions which science has been able to discover]

For example, the production of the first animals is announced in the words: 'God said, Let the waters swarm with swarmers.'[1] A naturalist here recognises not only the origination of animal life in the waters, but also three powers or agencies concerned in its introduction, or rather, perhaps, one power and two conditions of its exercise. First, there are the Divine power and volition contained in the words, 'God said.' Secondly, there is a medium or environment previously prepared and essential to the production of the result —'the waters.' Thirdly, there is the element of vital continuity in the term 'swarmers'—that reproductive element which hands down the organism with all its

[1] This is, perhaps, the best word to express the meaning of the term *sheretzim*—rapidly multiplying or dividing creatures.

B

powers from generation to generation, from age to
age. If we ask modern science what are the agencies
and conditions implied in the introduction on the
earth of the multitudinous forms of humble marine
life which we find in the oldest rocks, its answer is in
no essential respect different. It says that these
creatures, endowed with powers of reproduction and
possibly of variation, increased and multiplied and
filled the waters with varied forms of life ; in other
words, they were *sheretzim*, or swarmers. It further
says that their oceanic environment supplied the external conditions of their introduction and continuance, and all the varieties of station suited to their
various forms—'the waters brought them forth.'
Lastly, since biology cannot show any secondary
cause adequate to produce out of dead matter even
the humblest of these swarmers, it must here either
confess its ignorance, and say that it knows nothing
of such ' abiogenesis,' [1] or must fall back on the old
formula, ' God said.'

Let it be further observed that creation or making,
as thus stated in the Bible, is not of the nature of what
some are pleased to call an arbitrary intervention and
miraculous interference with the course of nature. It
leaves quite open the inquiry how much of the vital

---

[1] It is sometimes urged against the idea of creation that it implies
abiogenesis or production without previous life.   But there must have
been abiogenesis at some time, and probably more than once, else no
living thing could have existed.

phenomena which we perceive may be due to the absolute creative fiat, to the prepared environment, or the reproductive power. The creative work is itself a part of Divine law, and this in a three-fold aspect : First, the law of the Divine will or purpose ; second, the laws impressed on the medium or environment third, the laws of the organism itself, and of its continuous multiplication, either with or without modifications.

While the Bible does not commit itself to any hypotheses of evolution, it does not exclude these up to a certain point. It even intimates in the varying formulæ, ' created,' ' made,' ' formed,' caused to ' bring forth,' that different kinds of living beings may have been introduced in different ways, only one of which is entitled to be designated by the higher term ' create.' The scientific evolutionist may, for instance, ask whether different species, when introduced, may not under the influence of environment change in process of time, or by sudden transitions, into new forms not distinguishable by us from original products of creation. Such questions may never admit of any certain or final solution, but they resemble in their nature those of the chemist, when he asks how many of the kinds of matter are compounds produced by the union of simple substances, and how many are elementary, and can be no further decomposed. If the chemist has to recognise, say, seventy substances as elementary, these are to him manufactured articles, products of creation. If he should be able to reduce them to a much smaller number, even ultimately to only one kind of matter, he would not by such discovery be enabled to dispense with a Creator, but would only have penetrated a little more deeply into His methods of procedure. The biological question is, no doubt, much more intricate and difficult than the chemical, but is of the same general character. On the prin- ciples of Biblical theism, it may be stated in this way: God has created all living beings according to their kinds or species, but with capacities for variation and change under the laws which He has enacted for them. Can we ascertain any of the methods of such creation or making, and can we know how many of the forms which we have been in the habit of naming as distinct species coincide with His creative species, and how many are really results of their variations under the laws of reproduction and heredity, and the influence of their surroundings?

I may add that this introductory chapter is neces- sarily a very general summary of the questions to which it relates, and that its positions will be much strengthened by our detailed consideration of those marvellous structures and functions of animals and plants which modern science has revealed to us, and their wonderful history in geological time. These are facts so stupendous in their intricacy and vastness that they make the relation of God to the origination and history of any humble animal or plant as grand

and inscrutable as His relation to the construction of the starry universe itself.

It is plainly shown by recent controversies, as, for example, those which have appeared in the *Nineteenth Century* for 1889, that the agnostic evolution and the acceptance of the results of German criticism in disintegrating the earlier books of the Bible, are at the moment combining their forces in the attack on Evangelical Christianity. They present a very formidable front, but if met in a spirit at once fair and firm, and with an intelligent knowledge of nature and revelation, the evil which they may do will be only temporary, and may lead in the future to a more robust and enlightened faith.

CHAPTER II

WHAT IS EVOLUTION?

IT is quite necessary to ask this question, since under the name Evolution so many things are vaguely included that, without care, we may involve ourselves in mental confusion.

1. Evolution sometimes professes to explain the origins of things ; but of this it knows absolutely nothing. Evolution can take place only where there is something to be evolved, and something out of which it can be evolved, with adequate causes for the evolution. This is admitted in terms by Darwin and his followers, but constantly overlooked in their reasoning, in which evolution is spoken of as if it were, or could be, an efficient cause. The title *Origin of Species* was itself a misnomer as used by Darwin. The book treated not of the origin of species, but of the transmutations of species already in existence.

2. The term Evolution as popularly used may thus include processes either modal or causal. The former implies development under adequate causes, and this is rational evolution. The latter assumes

to be itself a cause, which is in the nature of things impossible. The causes of development must always be distinguished from the evolution itself. It has been the fashion to use the expression ' factors of evolution ' to cover the causes ; but it would be more honest to admit at once that there must be efficient and adequate causes for every development.

3. The term Evolution is used to express indifferently all changes of the nature of development, however different in kind from each other. Spencer's definition that evolution is the ' transformation of the homogeneous through successive differentiations into the heterogeneous ' would cover creation as well as development, in the sense in which he understands it, and it does not cover those developments in which the complex becomes more simple, as in what is termed retrograde development in plants and animals. But this definition covers, as used by Spencer and Darwin, even with reference to organisms alone, three distinct things : (1) Direct development of structures previously prepared and subjected to the action of adequate causes, as heat, moisture, air, &c. Of this kind is the development of seeds and eggs into perfect plants and animals. This is the only kind which can be termed spontaneous, and this term can be applied only in a very limited sense, because it implies a previous laying up, potentially or structurally, in the germ, of all that is to be developed from it. (2) Indirect development, or that which takes place under

the power and guidance of an external will. Such is
the production of varieties of animals and plants by
selection and other means, and such would be creation
if carried out by a Supreme Being using His own
materials and laws. This, be it observed, is the only
sense in which there can be such a thing as natural
selection. Nature is either a purely imaginary being,
a mere figure of speech, or another name for a creative
will. (3) The supposed development of new kinds
or species of animals and plants from others by descent,
with modification—a process as yet unknown except
hypothetically and inferentially, and which is what
the doctrine of evolution is contrived to establish, in
so far as specific types are concerned, though it is
well known in the case of mere varieties. (4) The
supposed evolution of living organisms from dead
matter, also a process unknown to science—a creative
fact, which must have occurred at some time, but of
the nature and secondary causes of which we know
nothing. We may be certain, however, that if it was
in any sense of the nature of a development, this
must have been different from anything known to us
as occurring at present.

All these entirely distinct kinds of change are
mixed up by evolutionists in treating of organic evolution ; and they freely extend the same term to things
so different as the physical changes by which the
earth assumed its present form, the improvement of
arts and social institutions, the growth of nations by

human agency, and even the supposed development of the mind of man himself from the powers of lower animals. In these circumstances, if we are to understand anything of this confused and multiform philosophy, we must perpetually question its advocates and exponents as to the kind of development of which they are speaking, and as to the causes to which such alleged development may be attributed. We must also be especially cautious in scrutinising any analogies presented to us, as, for example, that between the development of an embryo into a perfect animal, and the succession of animals in geological time. In such a case we must inquire not only if the alleged developments are really similar, but if they take place in similar conditions and under the influence of similar causes—in other words, whether the analogy is real or only apparent.

So dangerous is this use of the term evolution, that it may become necessary to abandon the word altogether in purely scientific discussions, and to insist on the terms *causation* and *development*, as covering the two distinct ideas now mixed up under evolution. It is at least necessary in discussions on this subject to be constantly on our guard as to the kind of evolution in question, whether modal evolution of a direct or indirect, literal or figurative character, or the mere figment of a causal evolution.

With reference to the Darwinian system proper, this kind of definition is not difficult. Darwin's

natural turn of mind and his scientific training were
not of such a character as to lead him to seek for ultimate causes.   He was content with a modal evolution.
He took matter and force and their existing laws as
he found them.   He presupposed also life and organisation with all their powers, and even seemed to
postulate certain species of animals and plants as
necessary raw material wherewith to begin his process of evolution.   How all this vast and complex
machinery came into being he did not concern himself, and was content to leave it as something beyond
his ken.   Thus, as it appears in the *Origin of Species*,
evolution is merely a modification of specific forms,
and Darwin was content to explain this by an imaginary struggle for existence, and a supposed natural
or spontaneous selection exercised in an indefinite
way by external forces and conditions.   Thus it really
did not touch the question of how the first species
originated, but only that of their subsequent modification ' by means of natural selection,' or ' preservation
of favoured races in the struggle of life.'

Darwin thus did not concern himself much with
causal evolution, or the origin of things properly so-
called.   Indeed, when questioned on these points, he
appears to the last to have been in uncertainty and to
have desired not to commit himself.   To men whose
minds are not under the influence of positive theism,
or of a belief in Divine revelation, and who attain to
large acquaintance with nature, it either resolves it-

self into a cosmos which manifests the power and divinity of a creative will, or it becomes disintegrated into a chaos of confused and conflicting forces battling with one another. Darwin's view was of the latter kind, and hence to him the life of organised beings was a struggle for existence, or, at least, this appeared to him far more potent than the opportunity and desire to improve and advance, on which the great French naturalist, Lamarck, based his theory of evolution.

It is evident that such a view of nature has the appearance, at first sight, of being wholly subjective and illusory. It does not touch the question of origins. It assigns no adequate causes for either the movement or the uniform direction of the supposed development. It seems to enthrone chance or accident or necessity as Lord and Creator, and to reduce the universe to a mere drift, in which we are embarked as in a ship without captain, crew, rudder, or compass, and without any guiding chart or star.

Let us inquire, however, how Darwin justified a position apparently so unscientific. He took his initial stand on the idea that, as he expresses it, ' a careful study of domesticated animals and plants would offer the best chance of making out this obscure problem ' of the introduction of new species. Hence he was led to study the variation of animals and plants under domestication, and to infer similar effects as taking place in nature by a spontaneous power of ' natural selection ' exercised by the environment.

Thus, by a striking inversion of ordinary probabilities, inanimate nature was made to rule, determine, and elevate that which lives and wills. Singular though it may appear, this apparent paradox is one of the great charms of the doctrine to the general mind, which is excited by the strange and marvellous, especially when this is supposed to be countenanced by science.

This leading idea Darwin supported by several collateral considerations, such as the ascertained succession of animal and vegetable life in geological time, the analogy with this of the stages of the embryo in its development in the higher animals, the supposed power of sexual selection and the influence of geographical distribution. All these influences, including natural selection, were supposed to operate in a very slow and gradual manner, so much so that the observation of the apparent permanence of species within the human period should not be regarded as an objection.

The Darwinian system thus embraced a modal evolution or development of living beings, with certain alleged causes keeping up the movement and giving it direction ; and all this with or without a superintending will and creative power behind it. Presented in an attractive and popular manner, and with a great mass of facts supposed to sustain it, and concurring with the popular evolutionary philosophy of Herbert Spencer, it was at once accepted by a great number of scientific and literary men, and applied in varied

ways to the solution of many questions more or less
analogous to that of the origin of species, while, as was
natural, it has been pushed in a vast number of wild
and extreme directions by popular writers not con-
versant with science in a practical manner. It has,
however, been seriously canvassed by the more
cautious and conservative men of science, and has
been found to fit in so badly with what is actually
known of nature, that it has gradually been obliged to
modify its claims ; and ultimately its adherents have
become divided into distinct schools, differing materially from each other and from the original Darwinism,
though all agree in claiming Darwin as a master and
in upholding his merit as a great discoverer. These
various schools are divided : (1) As to the primary
causes of the development ; (2) As to the secondary
causes ; (3) As to the mode or modes.

With reference to the first, there are some evolutionists who are agnostic like Spencer, monistic like
Haeckel, or merely negatively materialistic, like a
large number of the younger naturalists. On the other
hand, there are advocates of evolution who profess to
see in it the manifestation of Divine creative power,
and with whom evolution is merely the manner in
which the will of God manifests itself.

With reference to the secondary causes supposed
to be at work, observation and experiment have
shown that, if development of new species has taken
place, other causes than those alleged by Darwin

must have been operative, although many able Darwinians, like Weismann and Wallace, profess to regard
natural selection as the sole operative cause. The
influence of an innate tendency to vary has been
claimed by some, as if in the original creation of
living beings they had been so wound up as to
go first in one direction and then in another,
without any external cause, or when acted upon
by varying causes. The influence of favourable
conditions and room for expansion has been alleged
by others, in accordance with the old view of Lamarck.
The tendency of some lower animals, under unfavourable conditions, to become reproductive before they
have attained to full maturity, while more favourable
circumstances elevate the standing to which the
animal attains before producing young, is also a consideration which, under the name of reproductive
acceleration or retardation, has attracted some attention. Various causes of abrupt or sudden change
have also been invoked, as, for instance, those obscure
agencies which determine the appearance of monstro-
sities or varietal forms among domesticated animals.

The question of efficient cause has thus become
very complicated, and the only points on which all
are united are the possibility of varieties or races in
some way overleaping the bounds of specific fixity
and becoming new species, and the further doctrine
that changes acquired in any way may become permanent as an inheritance in such new species. This

last tenet of heredity has, however, of late been greatly shaken by the investigations of Weismann, which have thrown doubt on the possibility of inheritance of some characters acquired by the individual. We shall see that if these new views are established, the whole aspect of the question of specific modification will be greatly changed. Since, however, no case establishing any one of the alleged factors of new species is actually known to have occurred, these doctrines of modification and heredity, as applied to the origin of species, are, as yet, articles of faith and not of scientific certainty, and the whole question of causation in evolution may be said to be in an uncertain and transition state.

In these circumstances the questions as to possible modes of development may seem to lose much of their importance ; but the disciples of Darwin inform us that, independently of known and ascertained causes, the probability of development which arises from embryonic analogy and the affinities of animals and plants among themselves, is so great that the doctrine must nevertheless be credited or at least treated with respect. Farther, the modes of development are, as we have already seen, the only points on which certain evidence can be obtained. It is necessary, therefore, to consider these.

Here we must admit, in the first place, that though we can study modes of variation of species, no case has actually occurred under the observation

of naturalists of the development of a new species.
We must also admit that such is the fixity of specific
forms at present, and the nice equilibrium of all their
parts, that the changes effected under domestication
and by artificial selection seriously unsettle their
stability, and cause the varieties and races produced
to exist under a condition of tension and unstable
balance, which renders them infertile and otherwise
unlikely to survive if left to themselves. They have,
farther, in favourable circumstances a strong tendency
to revert to the original types. Again, we must admit
that on the supposition of slow and piecemeal altera-
tion in a complex organism, we meet with endless
difficulties in relation to the origin of each change,
its fitting in with the other parts of the organism and
its maintenance while still too imperfect to be of use.
These difficulties are specially formidable when the
whole depends on favouring accidents in the absence
of a guiding will like that of the human breeder. We
also find that in the past history of life in geological
time, there are several great difficulties in the way of
the idea of slow and gradual modification.

One arises from the fact that we can trace most of
the leading types so far back that they seem to constitute parallel rather than diverging lines, and show
no certain evidence of branching. The continuance
of the Lingulæ and other Brachiopods, and of the
silicious sponges and the Foraminifera, from the
Cambrian to the modern, and more lately the history

of the oysters, which have continued from the Carboniferous age to the present, and that of the scorpions,
which have continued from the Silurian, in both cases
with scarcely any more differences than their successors present at the present day, may be taken as
examples. With this must be connected the further
fact that nearly all the early types of life seem very
long ago to have reached stages so definite and fixed
that they became apparently incapable of further
development, constituting what have recently been
called 'terminal forms.'[1]

A further difficulty arises from our failure to find
satisfactory examples of the almost infinite alleged
connecting links which must have occurred in a
gradual development. This, it may be said, proceeds
from the imperfection of the record; but when we
find abundance of examples of the young and old of
many fossil species, and can trace them through their
ordinary embryonic development, why should we not
find examples of the links which bound the species·
together? An additional difficulty is caused by the
fact that in most types we find a great number of
kinds in their earlier geological history, and that they
dwindle rather than increase as they go onward. This
fact, established in so many cases as to constitute an
actual law of palæontology, is altogether independent
of the alleged imperfection of the record.

Objections of this kind appear to be fatal to the

[1] Clelland, *Journal of Anatomy and Physiology.*

Darwinian idea of slow modifications, proceeding throughout geological time, and to throw us back on a doctrine of sudden appearance of new forms, occurring at certain portions of geological time rather than at others, and in the earlier history of animal and vegetable types rather than in their later history, and in early geological times, rather than in those more recent. This doctrine, however, of critical or spasmodic evolution is essentially different from Darwinism, and approaches to that which has been called mediate creation, or creation under natural law.

With respect to the origin of man himself, which is, no doubt, the most important point to us, these difficulties are enormous. We can trace man only a little way back in geological history, not farther than the Pleistocene period, and the earliest men are still men in all essential points, and separated from other animals, recent and fossil, by a gap as wide as that which exists now. Farther, if from the Pleistocene to the modern period man has continued essentially the same, this, on the principle of gradual development, would remove his first appearance not only far beyond the existence of any remains of man or his works, but beyond the time when any animals nearly approaching to him are known to have existed. This is independent altogether of the farther difficulties which attend the spontaneous origination of the mental and moral nature of our species. It would

seem, then, that man must have been introduced, not
by a process of gradual development, but in some
abrupt and sudden way. Even Wallace, who has all
along adhered to the doctrine of natural selection in
its integrity, while he agrees with Darwin that man
must be a descendant of apes as to his bodily frame,[1]
maintains that his higher mental and moral faculties
must have had another origin.

These considerations have led many of the more
logical and thoughtful of the followers of Darwin to
the position of supposing, not a gradual, but an intermittent and sudden development, and this, in the
main, in the earliest periods of the history of living
beings. In a very able essay by Dr. Alpheus Hyatt,
in the *Proceedings* of the Boston Society of Natural
History, this view is very fully stated in its application to animals. On the one hand, Hyatt holds that
the biological facts and the geological evidence as it
has been stated by Marcou, Le Conte, Barrande,
Davidson, and by the author of this work, precludes
the idea of slow and uniform change proceeding
throughout geological time, and he holds justly that
the idea of what he calls 'a concentrated and accele-
rated process of evolution,' in early geological times,
brings the doctrine of development nearer to the position of those great naturalists like Cuvier, Louis
Agassiz, and Gegenbaur, who have denied any genetic
connection between the leading animal types. He

[1] *Darwinism*, p. 461.

quotes Cope and Packard in support of his view on
this point. The latter we shall have occasion to refer
to in the sequel in connection with cave animals.
Cope has, in a series of brilliant essays,[1] endeavoured
to illustrate what he terms ' causes of the origin of the
fittest.' Of this kind are growth-force modified by
retardation or acceleration of development produced
by unfavourable or favouring conditions, the effects of
use and disuse on modifying structures, the law of
correlation of parts and the effects of animal intelligence. These are all causes ignored by the genuine
Darwinian. Nevertheless they exist in nature, though
rather as causes of mere adaptive variation than of
specific difference.

Another modification of orthodox Darwinism is
that of Romanes, who may almost be regarded as
Darwin's most prominent successor. He has introduced the idea of physiological selection, that is, of
the occurrence accidentally or from unknown causes
of reproductive changes which render certain individuals of a species infertile with others. The effect
of this would be an isolation amounting to the erection
of two forms not reproductive with each other ; or, in
other words, of two species not gradually differentiated,
but distinct from the first. This is really an inversion
of Darwin's theory, in which the initial stage of
Romanes is necessarily the culmination of the development. It differs also essentially in eliminating the

[1] ' Origin of the Fittest,' *American Naturalist.*

idea of use and adaptation to change implied in the theory of natural selection.

Romanes even goes so far as to stigmatise the adherence to natural selection pure and simple as 'Wallaceism,' in contradistinction to Darwinism, while he admits that Wallace has a good right to adhere to this view, as having in some sense antedated Darwin in asserting the dominant influence of natural selection. It is fair to say, with regard to Romanes, that while advocating the importance of 'Physiological Selection,' he claims that Darwin admitted, or would have admitted, this factor, since he believed that in the absence of infertility to prevent intercrossing, natural selection would fail to produce new species. It is worthy of remark here that both Romanes and Wallace seem to be aware that this admission might be fatal to the doctrine of natural selection, unless they can show some other cause capable of producing infertility.

In the meantime, Weismann in Germany has, in the name of what has been called pure Darwinism, introduced into the discussion facts and considerations as destructive to the usual doctrine as Puritanism would be to High Churchism. He contends that all evidence is against the perpetuation by heredity of characters acquired by the individual. Only characters born with him can be perpetuated. For example, a man born with six fingers on his hand may have six-fingered children, but a man who acquires in his lifetime manual dexterity, or who loses a finger by accident, will not transmit either peculiarity. Weismann has undoubtedly made out a strong case in favour of this contention, which would at once overthrow the Lamarckian theory of evolution, and would remove one of the subsidiary props of Darwinism, throwing it back entirely on the natural selection of fortuitous congenital variations. Purified in this way, and reduced to chance variation, perpetuated by accidental action of favouring circumstances, Darwinism would, according to some of its adherents, evaporate without leaving any residuum. Nor has it escaped notice that the theory of Weismann implies profound and far-reaching considerations respecting the independence of the germinal matter of animals of individual peculiarities, and its constancy to the ideal plan of the species, which would help us to account for the wonderful permanence of types in geological time, while it would oppose change, except when this arises from causes directly affecting the reproductive function.

Another important point involved in Weismann's results is the probability that, while asexual reproduction, as, for instance, that of budding, tends to perpetuate individual peculiarities, whether of advance or retrogression, ordinary reproduction tends to eliminate all variations, whether produced by habit and use or by obscure causes affecting the individual in its lifetime. Thus there is a strong barrier set up, especially in the higher organisms, against either degradation or elevation.

Advantage has been taken of this by some specu-lators to suggest that new species may have originated by parthenogenesis, that is to say, by what theologians would call miraculous conception, and this idea has by some of them been connected even with the nativity of our Lord on the earth. But such speculations are very far removed from even the borders of science. These speculations may, however, raise the question whether man is to be succeeded by any improved species. If it had pleased God at any time to produce several individuals of a new race as superior to ordi-nary humanity as was Jesus of Nazareth, and to isolate and protect from admixture this new departure, the world might have entered on a new stage as superior to the present as man himself is to the pre-daceous beasts which the nations of the earth delight to use as their emblems. This idea presented itself to the Prophet Daniel when he saw the successive conquering empires of the world represented by a series of ferocious beasts, and saw these replaced by one 'like unto the Son of Man,' a truly human per-sonage, descending from heaven to reign on earth. The same figure is in the mind of Christ when He calls Himself distinctively the 'Son of Man,' not as merely human or in comparison with God, but as contrasted with the lower powers of earth, and as re-presenting the heaven-descended man of Daniel.

Jesus, however, assures us that not a new species of *homo*, but man himself, in a redeemed, sanctified, and spiritual state, is to be the heir of the coming ages.

A curious point, little thought of by most evolutionists, but deserving consideration here, is that to which Herbert Spencer has given the name ' direct equilibration,' or the balance of parts and forces within the organism itself. The body of an animal, for example, is a very complex machine, and if its parts have been put together by chance, and are drifting onwards on the path of evolution, there must necessarily be a continual struggle going on between the different organs and functions of the body, each tending to swallow up the other, and each struggling for its own existence. This resolution of the body of any animal into a house divided against itself, is at first sight so revolting to common sense, and so hideous to right feeling, that few like to contemplate it ; but it has been brought into prominence by Roux and other recent writers, especially in Germany, and it is no doubt a necessary outcome of the evolutionary idea. For why should not the struggle of species against species extend to the individuals and the parts of the individual ? On this view, the mechanism of an animal ceases even to be a machine, and becomes a mere mass of conflicting parts thrown together at random, and depending for its continued existence on a chance balance of external forces. It is well for us that we have not in human machinery to deal with such unstable and dangerous combinations, else no one's life would be for a moment safe.

Fortunately, geological history so completely negatives this idea, by showing the extreme permanence of many forms of life which have continued to propagate themselves through almost immeasurable ages and great changes of environment, without material variation, and the apparent fixity of these in their final forms, that we are relieved from the dread which this nightmare of German brains tends to create.

Viewed rightly, the direct equilibration of the parts of animals and plants is so perfect and so stable, and such great evils arise from the slightest disturbance of it by the selective agency of man, that it becomes one of the strongest arguments against the production of new species by variation. This has been well shown by Mr. T. Warren O'Neill, of Philadelphia,[1] who adduces a great number of facts, detailed by Darwin himself, to show that when the stability of an organism is artificially altered by man in his attempt to establish new breeds, infertility and death of these varieties or breeds results ; and if this happens under the fortuitous selection supposed to occur in nature, any considerable variation would result either in speedy return to the original type or in speedy extinction. In other words, so beautifully balanced is the organism, that an excess or deficiency

---

[1] *Refutation of Darwin.* Philadelphia, 1880.

in any of its parts, when artificially or accidentally
introduced, soon proves fatal to its existence as a
species ; so that, unless nature is a vastly more skilful
breeder and fancier than man, the production of new
species by natural selection is an impossibility.

Two remarkable books by two of the ablest exponents of the Darwinian theory of evolution have
recently appeared, which may be taken as specimens of
the evolutionary method, and may be commended to
those who desire to know this theory as defended and
extended by its friends.[1]   One of these works is by
Alfred Wallace, who may be truly said to have anti-
cipated Darwin in the theory of natural selection—
the other by Dr. Romanes, Darwin's successor.   Both
claim to be orthodox Darwinians, though each accuses
the other of some heresy.   Wallace's book may, however, be accepted as the best English exposition of
Darwinism in general, that of Romanes as the ablest
attempt to explain on this theory the evolution of the
higher faculties of man.   Neither professes to explain
the origin of life, but both profess, life and species of
animals being given, to explain their development
as high as man himself, though they differ materially
as to this highest stage of evolution, and also as to
the omnipotence of natural selection.   The judicious
reader will, however, observe that both take for granted
what should be proved ; in other words, reason con-

---

[1] *Darwinism*, by Wallace ; *Mental Evolution in Man*, by
Romanes.

stantly in a narrow circle, and constantly use such formulæ as 'we may well suppose,' instead of argument.

We may take as an example from Wallace the history of the evolution of the water-ouzel or dipper. It may serve as an example of the questions which are raised by the Darwinian evolution, and which, if they have no other advantage, tend to promote the minute observation of nature, of which Wallace's book shows many interesting examples. It serves, at the same time, to illustrate that peculiar style of reasoning in a circle which is characteristic of this school of thought. I have chosen this special illustration from Wallace because it is one in which the idea of adaptation to fill a vacant space—an idea as much Lamarckian as Darwinian—is introduced.

An excellent example of how a limited group of species has been able to maintain itself by adaptation to one of these 'vacant places' in nature is afforded by the curious little birds called dippers or water-ouzels, forming the genus *Cinclus* of the family *Cinclidæ* of naturalists. These birds are something like small thrushes, with very short wings and tail and very dense plumage. They frequent, exclusively, mountain torrents in the northern hemisphere, and obtain their food entirely in the water, consisting, as it does, of water-beetles, caddis-worms, and other insect larvæ, as well as numerous small fresh-water shells. These birds, although not far removed in structure from thrushes and wrens, have the extraordinary power of flying under water ; for such, according to the best observers, is their process of diving in search of their prey, their dense and somewhat fibrous

plumage retaining so much air that the water is prevented
from touching their bodies, or even from wetting their feathers
to any great extent. Their powerful feet and long curved
claws enable them to hold on to stones at the bottom, and
thus to retain their position while picking up insects, shells,
&c. As they frequent chiefly the most rapid and boisterous
torrents, among rocks, waterfalls, and huge boulders, the
water is never frozen over, and they are thus able to live
during the severest winters. Only a very few species of
dipper are known, all those of the old world being so
closely allied to our British bird that some ornithologists
consider them to be merely local races of one species ;
while in North America and the Northern Andes there are
two other species.

I here, then, we have a bird, which, in its whole structure,
shows a close affinity to the smaller typical perching birds,
but which has departed from all its allies in its habits and
mode of life, and has secured for itself a place in nature
where it has few competitors and few enemies. We may
well suppose that, at some remote period, a bird which was
perhaps the common and more generalised ancestor of most
of our thrushes, warblers, wrens, &c. had spread widely
over the great northern continent, and had given rise to
numerous varieties adapted to special conditions of life.
Among these some took to feeding on the borders of clear
streams, picking out such larvæ and molluscs as they could
reach in shallow water. When food became scarce they
would attempt to pick them out of deeper and deeper water,
and while doing this in cold weather many would become
frozen and starved. But any which possessed denser and
more heavy plumage than usual, which was able to keep
out the water, would survive ; and thus a race would be
formed which would depend more and more on this kind of
food. Then, following up the frozen streams into the
mountains, they would be able to live there during winter ;

and as such places afforded them much protection from enemies and ample shelter for their nests and young, further adaptations would occur, till the wonderful power of diving and flying under water was acquired by a true land-bird.[1]

Here it will be seen that a bird, distinctly marked off by important structures and habits from others, is supposed to have originated from a different species at some remote period, by efforts to obtain food in what, to it, must have been an unnatural way ; and the sole proof of this is the expression, 'we may well suppose.' Why may we not as well suppose that all the perching birds were at first like water-ouzels, which would accord with the early appearance of aquatic birds, and that they gained their diverse forms by availing themselves of the better circumstances and more varied food to be found in the woods and fields, so that our water-ouzel may be a survival of a primitive type? Neither theory can be proved, and the one is as likely as the other, perhaps the latter, of the two, the more likely, and neither actually explains anything. It is to be observed, also, as already hinted, that the kind of evolution in this, as in some other cases supposed by Wallace, is rather Lamarckian than Darwinian.

It is interesting to note that, though wedded to that strange mode of reasoning of which the extract above given furnishes an example, Wallace frankly and fully admits three of the great breaks in the con-

[1] *Darwinism*, pp. 116, 117.

tinuity of evolution. First, he admits that we cannot
account for the introduction of life at first, because we
know no way in which mere chemical combination
can produce living protoplasm. Here, he says, ' we
have indications of a new power at work which we
may call Vitality.' Secondly, he sees no cause in the
continuous evolution for the introduction of animal
sensation and consciousness. No attempt at explanation by any modification of protoplasm can here
' afford any mental satisfaction, or help us in any way
to a solution of the mystery.' He sees a similar
break of continuity in the introduction of the higher
faculties of man. ' These faculties could not have
been developed by means of the same laws which
have determined the progressive development of the
organic world in general and also of man's physical
organism.' These he refers to an unseen universe—
to a world of spirit to which the world of matter is
altogether subordinate. If we refer these three great
steps to a spiritual Creator, and eliminate, on the other
side, the known development of varietal forms, the
field for the Darwinian evolution becomes greatly
narrowed.

Romanes, the author of the other work, will listen
to no such compromises ; but, on the other hand, is
willing to admit a union of the Darwinian and
Lamarckian doctrines, besides sexual selection and
other factors, which are admitted also by Spencer.
His latest work is devoted to the bridging over the

third of the gaps above mentioned, as in a previous
work he had dealt with the second. He does not
affirm that he has fully succeeded, but that, by considering the case of savages and of prehistoric man,
we 'are brought far on the way towards bridging the
psychological distance which separates the gorilla
from the gentleman.' It is one thing, however, to be
on the way to a chasm, and another to be assured that
there is a good bridge over it. If we succeed in crossing with him from instinct to animal intelligence, from
this to rational thought, from this to ethical judg-
ments and to the belief in God and immortality, and
along with all this to speech, we have the following
to reward us in regard to one step of our progress :
' I believe that this most interesting creature (speechless man) lived for an inconceivably long time before
his faculty of articulate sign-making had developed
sufficiently far to begin to starve out the more primitive and more natural systems ; and I believe that
even after this starving-out process did begin, another
inconceivable lapse of time must have been required
to have eventually transformed *Homo alalus* into
Homo sapiens.' A process which thus requires two
eternities in which to pass through two of its stages
may well stagger the credulity of ordinary specimens
of *Homo sapiens*, and may surely be dismissed as
itself ' inconceivable.'

While, however, the conclusions of Romanes are
thus somewhat unsatisfactory, his book contains much

that is valuable, more especially with reference to the perfectly legitimate questions relating to the development of civilisation, and of new ideas and inventions in human history. Man is not confined, like the lower animals, within the range of unvarying instinct. He is gifted with inventive and progressive powers, and in the study of the progress of these there is scope for much psychological inquiry and discussion, though it is evident that human progress is not of the nature of a slow and gradual evolution, but rather by sudden leaps under the influence of superior genius and mental power, and it is all within the specific limits of man, and in no respect tends to the production of a new species.

This general view of evolution will enable us to have some definite idea of the doctrine as presented by Darwin and his followers ; but perhaps it may be well before proceeding farther to consider what bearing it may have on theology, or more properly whether it accords with or contradicts the idea of divine creation as maintained in Revelation. As to this, little apprehension need be entertained on the part of Christianity, and it may safely leave such questions as those above discussed to exhaust themselves, except in so far as they may affect the interest of individual unstable souls. This last is, however, an important matter, and it may be well to scrutinise it more closely.

The modern hypotheses of evolution present themselves to the Christian under two aspects—the theistic and the atheistic or agnostic, for the two last are practically the same. The theistic evolutionist holds that God creates, but that created things may have powers of spontaneous evolution, under laws whereby they may pass into new and higher forms. The atheist and the agnostic eliminate the idea of a Creator, and reduce everything to the action of atoms and forces supposed to be practically and inherently omnipotent. They thus make of these atoms and forces a supreme god, attributing to them the same powers assigned by the theist to the Creator. It is obvious, however, that many adherents of evolution have no clear perception of the distinction between these phases, or find it convenient to overlook its existence, since we often find them hovering in thought between the one and the other, or occupying one or the other position indifferently, as the exigencies of debate may require.

It is also to be observed that either of these phases of evolution may admit of modifications. One of the most important of these arises from the distinction between the idea of slow and uniform development maintained by Darwin and others, and that of sudden or intermittent evolution advocated by such evolutionists as Mivart and Le Conte.

Viewing the matter in this light, it is evident that neither the theological idea of creation nor the evolutionist notion, in either of its phases, can have any

D

close dependence on biological and geological science, which studies the nature and succession of organic forms without ascertaining their origin ; either hypothesis, may, however, appeal to scientific facts as more or less according with the consequences which might be expected to follow from the origins supposed. It is further evident that, should evolutionists be driven by natural facts to admit the sudden apparition of organic forms rather than their gradual development, there may be no apparent difference, as to matter of fact, between such sudden apparition and creation, so that science may become absolutely silent on the question.

Palæontology has indeed recently tended to bring the matter into this position, as Barrande and others have well shown. [I have myself elsewhere adduced the advent of the Cambrian trilobites, of the Silurian cephalopods, of the Devonian fishes, of the Carboniferous batrachians, land snails and myriapods, of the marsupial mammals of the Mesozoic and the placental mammals of the Eocene, and of the Palæozoic and modern floras, as illustrations of the sudden swarming in of forms of life over the world, in a manner indicating flows and ebbs of the creative action, inconsistent with Darwinian uniformity, and perhaps unfavourable to any form of evolution ordinarily held.]

¹ In England, Davidson, Jeffreys, Williamson, Carruthers, and other eminent naturalists have strongly insisted on the tendency of palæontological facts to prove permanence of type and intermittent

This neutral attitude of science has been strongly insisted on by Dr. Wigand[1] in his elaborate work *Darwinismus*, in which he holds that this doctrine does not represent a definite and consistent scientific effort and result, but merely an ' indefinite and confused movement of the mind of the age,' and that science may ultimately prove its most dangerous foe. In like manner the veteran German physiologist Virchow, in an able address before the Assembly of German Naturalists at Munich,[2] taking the spontaneous generation of organisms and the descent of man from ape-like ancestors as test questions, argues in the most conclusive manner that neither can be held as a result of scientific investigation, but that both must be regarded as problems as yet unsolved.

But in the face of such opinions as these, we are struck with the fact that eminent men of science in England and America inform us that science demands our belief in the theory of evolution, and this in its atheistic as well as its theistic phase. When, however, we ask reasons for this demand, we find that those who make it are themselves obliged to admit the absence of a scientific basis for the doctrine.

For example, I may refer to the able and elaborate address delivered a few years ago before the American Association by its President, Professor Marsh. He

introduction of new forms, as distinguished from descent with gradual modification.

[1] Dr. Albert Wigand, *Darwinismus*, 1875-7.
[2] On the ' Liberty of Science,' 1877.

says : ' I need offer no argument for evolution, since
to doubt evolution is to doubt science, and science is
only another name for truth.' In the sequel of the
address he limits himself to the evolution of the vertebrate animals, admitting that he knows nothing of
the absolute origin of the first of them, and basing his
conclusions mainly on the succession, in distant times,
and often in distant places, of forms allied to each
other, and advancing in the scale of complexity.
Such succession obviously falls far short of scientific
proof of evolution ; and other than this no evidence
is offered for the strong assertion above quoted. In
the conclusion of the address he asserts that life may
be a form of some other force, presumably physical
force ; but admits in the same breath that we are
ignorant of its origin ; and finally he makes an appeal,
not to facts, but to faith : ' Possibly the great mystery
of life may thus be solved ; but whether it be or not,
a true faith in science knows no limit to its search for
truth.' Plainly, if this is all that can be said as to
scientific results concerning the origin of life, if this
origin is still an unsolved problem, a 'great mystery,'
it is a somewhat strong demand on our faith to ask
us to believe even that science will in the future succeed in effecting the solution of this problem, and we
should not have been told that to doubt evolution is to
doubt science. This style of treating the subject is
indeed much to be deprecated in the interest of science
itself.

Another eminent apostle of evolution, Professor Tyndall, tells us, in a recent public address, that 'it is now very generally admitted that the man of to-day is the child and product of incalculable antecedent time. His physical and intellectual textures have been woven for him through phases of history and forms of existence which lead the mind back to an abysmal past.' But, however generally this may be 'admitted,' it is nevertheless true that the oldest known men are as truly human in their structures as those now living, and that no link between them and lower animals is known. In a previous address he had gone further back still, and affirmed that in material atoms reside the 'promise and potency of life'; yet in his capacity of physicist he has by rigid experiments in his laboratory done as much as any man living to convince us that science knows no possibility of producing the phenomena of life from dead matter.

Perhaps no example could more vividly portray the contrast between exact science and evolutionist speculations than the careful experiments on germs suspended in the atmosphere made by Tyndall in the laboratory of the Royal Institution—experiments so complete, so convincing, and so eminently practical in their bearing on the conditions of health and disease—as compared with the quaint and crude imaginings of the same mind when, in the presence of popular audiences, it speculates on evolution.

But we should not too strongly denounce these speculative tendencies of scientific minds. They may point the way to new truths, and in any case they have an intense subjective interest. Nothing can be more interesting in a psychological point of view than to watch the manner in which some of the strongest and most subtle minds of our time exhaust their energies in the attempt to solve impenetrable mysteries, to force or pick the lock of natural secrets to which science has furnished no key. The objectionable feature of the case is the representation that such efforts have any real scientific basis.

Whence, then, arise these strange inconsistencies and contradictions which infest modern science like parasites? The expression I have already quoted is the only solution. They represent ' a confused movement of the mind of the age '—of an age strong in material discoveries, but weak in self-control and higher consciousness. The mind of our time is unsettled and restless. It has a vague impression that science has given it the power to solve all mysteries. It is intoxicated with its physical successes, and has no proper measure of its own powers. It craves a constant succession of exciting and sensational generalisations. Yet all this frenzy is no more the legitimate outcome of science than the many fantastic tricks which men play in the name of religion are the proper results of revelation or theology.

The true remedy for these evils is twofold. First,

to keep speculation in its proper place as distinct from science ; and secondly, to teach the known facts and principles of science widely, so that the general mind may bring its common-sense to bear on any hypothesis which may be suggested. Speculations as to origins may have some utility if they are held merely as provisional or suggestive hypotheses. They become mischievous when they are introduced into text-books and popular discourses, and are thus palmed off on the ignorant and unsuspecting for what they are not.

The man who, in a popular address or in a text-book, introduces the 'descent of species' as a proved result of science, to be used in framing classifications and in constructing theories, is leaving the firm ground of nature and taking up a position which ex-poses him to the suspicion of being a dupe or a charlatan.[1]   He is uttering counterfeits of nature's currency. It should not be left to theologians to expose him , for it is as much the interest of the honest worker in science to do this as it is that of the banker or merchant to expose the impostor who has forged another's signature.   In the true interests of

[1] I am glad to observe that Huxley, in the preface to the *Manual of the Anatomy of the Invertebrated Animals* (1878), has taken this ground.   He says : 'I have abstained from discussing questions of ætiology, not because I underestimate their importance, or am in-sensible to the interest of the great problem of evolution, but because, in my mind, the growing tendency to mix up ætiological speculations with morphological generalisations will, if unchecked, throw biology into confusion.'

science we are called on to follow the weighty advice of Virchow : ' Whoever speaks or writes for the public ought, in my opinion, doubly to examine just now how much of that which he says is objective truth. He ought to try as much as possible to have all in- ductive extensions which he makes, all conclusions arrived at by the laws of analogy, however probable they may seem, printed in small type under the general text, and to put into the latter only that which is objective truth.' To practise such teaching may require much self-denial, akin to that which the preacher must exercise who makes up his mind to forego his own thoughts, and, like Paul, to know no- thing among men but God's truth in its simplicity. The mischief which may be done to science by an opposite course is precisely similar to that which is done to religion by sensational preaching founded on distortions of scriptural truth, or on fragments of texts taken out of their connection and used as mottoes for streams of imaginative declamation.

To render such evils impossible, we must have a more general and truthful teaching of science. It is a great mistake here to suppose that a little knowledge is dangerous ; every grain of pure truth is precious, and will bear precious fruit. The danger lies in mis- using the little knowledge for purposes which it can- not serve ; and this is most likely to take place when facts are not known at all, or imperfectly compre- hended, or so taught as to cause a part of the truth to be taken for the whole. Let the structures of animals and plants in some of their more prominent forms be well known, along with their history in geological time, and the attempt to explain their origin by any crude and simple hypotheses like those now current will become unreal as a dream. These general statements must, however, be tested in the subsequent chapters by an appeal to facts in connection with the more important questions raised by modern evolution.

CHAPTER III

THE ORIGIN OF LIFE

IT has been remarked as a somewhat significant circumstance that the title of that remarkable work *The Origin of Species by Natural Selection*, which has so deeply impressed the mind of our age, contains in itself the elements of the refutation of its own leading principle.

Of the *origin* of species the book tells us nothing. It merely discusses certain possible modes of 'descent with modification' whereby new species may be derived from those previously existing. Of *species* it tells us nothing, except that if its contentions be maintained there can be no permanent kinds of animals or plants, or true species, in the old sense of the term, but only an indefinite shading of forms into one another, and a perpetual flux, by which what may be called a species at one period will be something different at another. *Natural selection* again, if there is such a thing, can take place only after species already exist with numerous individuals to be selected from ; and unless it is merely another name for chance, it implies also an intelligent selecting power or agency. Farther, though put forward by Darwin as an efficient cause, it is now admitted by many of the ablest of his followers[1] to be merely a mode, and only one of several modes by which species may be modified.

This error in statement proceeds from a fundamental confusion in the mind of the author, who, though of transcendent gifts as an observer, was very defective as a reasoner. He does not clearly discriminate between the origin, beginning, or development of living beings and the nature of the forms under which they present themselves in our more or less arbitrary systems of classification. This confusion pervades the whole work, and is accompanied by a like confusion between causation and development—ideas which are variously combined under the complex notion of evolution—so that such factors as struggle for existence and natural selection may be viewed sometimes as true, efficient powers and sometimes as mere modes or processes.

The initial question before us in this matter is: How did that which possesses organisation and life originate from that which is destitute of these properties? If we can answer this question, that of modification may follow in due course. If we cannot solve the problem of the origin of the life, the other question as to specific forms becomes of secondary

[1] Spencer, Romanes, Packard, Cope, &c.

significance. To those who are impressed with the necessity of an almighty creative will as the primary cause of all things, the formula 'God created' may be a sufficient answer; and it is perfectly true that we cannot expect the methods of science to go back to origins. It deals rather with laws and modes, and must necessarily always start from certain axioms, postulates, or powers of unknown origin. Still we must bear in mind that revelation itself invites us to think of such questions. It does not merely say that God created all things. It informs us that God said, 'Let the land bring forth plants, let the waters swarm with living things,' and it speaks of the bodily frame of man as made of the dust, that is, of the common material of the earth, and of an inspiration of the Almighty, an inbreathing of the breath of life to give him understanding.

These statements invite us from the side of revelation as well as of science to consider the antecedents and materials of life. ' In the beginning God created the heavens and the earth.' Here we have a fundamental statement which demands no proof, because we can substitute nothing else for it. If we say, ' There was no beginning, the Universe is eternal,' we have a proposition unthinkable by us, because we cannot imagine an eternal succession, and such a succession, if conceivable, would preclude all development. If we say, ' In the beginning the heavens and the earth were self-created,' we have a proposition which is a contradiction in terms. It remains as the only possible alternative that all things were created by the Almighty Intelligent Will whom we call God.

But having settled so much, we have presented to us a group of primary factors for the subsequent development. There are here, first, duration and extension—time and space, as we usually call them. We say we believe in time and space, because we know that we exist in them, but abstractly they are as inconceivable to us as the Being who exists from everlasting to everlasting, to whom one day is as a thousand years, and a thousand years as one day, whom the heaven of heavens cannot contain. Our definite ideas of these mysterious entities are based on things that extend and move. Time we may know by the succession of our own thoughts or the changes of external objects, space by the extension of the visible universe. Matter and motion are therefore our measures of extension and duration, and apart from these we may hold with Kant that time and space have no objective existence. But matter and motion must have had a beginning, and before that beginning time and space existed only in the Infinite mind, while to us the bodies that constitute the visible heavens are for times and for seasons, for days and for years; yet without time and space what remains to us except an immovable mathematical point? In regard to time and space, therefore, we may be agnostics if we please, for they are unknowable, but in connection with matter and motion we know them intimately, and cannot dissociate them from our thoughts. Practically, to us time and space begin with that beginning when God created the heavens and the earth.

We must next assume as factors in the development of the Divine plan a triad of things and powers existing in space and time—matter, ether, energy. What is matter? An aggregate of atoms, invisible and impalpable, yet believed in, and known to have different properties, dividing them into very dissimilar kinds or species, which, though related to each other with great regularity by definite gradations and numerical connections, cannot (so far as we know) be changed. They are species which seem to have no descent, and are not capable of modification. They are held together by equally inscrutable forces of affinity and cohesion, constituting what we call bodies, which, however solid, are permeable indefinitely by ethereal undulations.

Ether, again, is, so to speak, an immaterial matter, existing everywhere, yet incapable of perception—an inconceivable, all-pervading something, ministering to every sensation and action, yet itself imperceptible and inert. Next we have energy, manifesting itself by motion of different kinds, whether in ether or ordinary matter, and actuating all things, sometimes in one way, sometimes in another, yet under intelligible laws, which limit the freedom of energy and enable us to distinguish its different kinds or modes of manifestation.

All these things — space, time, matter, ether, energy—are to us inscrutable in their origin, and incapable of annihilation, yet in our power to deal with, according to certain laws which we have ascertained, and, no doubt, capable of endless changes and interactions as yet beyond our ken. Those that we do know constitute the subject matter of the vast and complicated sciences of physics and chemistry.

All this must have been present in the world, and as perfectly and regularly arranged as it is now, before there could be life. We may even say that all this must have been fully perfected, so as to admit of no farther improvement or change, before the origin of life. This is overlooked by those who unthinkingly tell us that we must believe in the evolution of the physical world, whether we believe in that of life or not. The development, in so far as the physical world is concerned, consisted in the arrangement and determination of matter and energy in such a manner as to fit the world for being the abode of life.

A vaporous world, a mere cloud or nebula of fire mist, a liquid incandescent world, a world with a hardened crust and a vaporous atmosphere, a world with a universal ocean covering its surface, a world with land and water, mountain and valley—all these may have existed (probably did exist) for untold ages before the origin of life. We know that in the earlier

of these stages the earth would be altogether unsuit-
able for any forms of life known to us ; but we do
not know precisely at what point of the later stages
it would be in precisely the best state for the begin-
nings of life. Let it be observed here, however, that
the materials of the physical world are to us manufactured or created products, and that the progress of
the development is the result of the properties and
laws impressed upon them at first, correlated and
regulated to a definite end. We shall find that there
is an analogy with this in the origin and development
of life.

But though all this material of the physical world
is necessary to life, it manifests in itself no indication
of that mysterious power ; for it we require something
more—namely, the substance protoplasm, which, so
far as we know, does not exist in dead nature, and
which thus far has baffled all attempts to construct it
artificially from its elements. In addition to this we
require some form of that complex machinery which
we call an organism, though this also, in our present
experience, cannot be formed without life. Yet
protoplasm and an organism must be present before
life can manifest itself.

Here we have another triad whose relations are
enshrouded in mystery. Just as we know nothing of
matter, ether, and energy, independently of each other,
so we know nothing of protoplasm, organism, and life,
except as existing together. We cannot imagine one

of them to originate without the other; nor can we imagine either of them to exist in nature in isolation from the others. All three are beyond our power to produce, and we have never witnessed their production spontaneously or by artificial means. Our inquiries so far have only brought us into the presence of two inscrutable and miraculous natural trinities. I say miraculous in the true sense of the term, because beyond our power and comprehension.

Protoplasm has been called the 'physical basis of life'; but this is merely a form of words to conceal ignorance. The substance is no doubt physical in the sense of being material and existing in nature, but it is not physical in the sense of being procurable or persistent under ordinary physical powers or conditions; and it is no more the basis of life and organisation than they are its basis. An egg is mainly composed of protoplasm—pure in the white, mixed with some other things in the yolk. It is also an example of dead or non-living protoplasm, though produced in the body of a living animal. But if fertilised it has in it a living and organised germ, also protoplasmic; and this germ can grow and assimilate the remainder of the protoplasm, and produce out of it all the parts of an animal even so complex as a bird. The animal so produced may have all the parts of a highly complex organic machine, made up of a number of special tissues, all of which were potentially, though not actually, present in the germ.

E

The protoplasm itself is a highly complex substance, consisting of carbon or charcoal, combined with three gases (oxygen, hydrogen, and nitrogen) and with minute quantities of sulphur and phosphorus, in molecules so complex that more than eight hundred atoms are supposed to be necessary to constitute one of them. But protoplasm alone immediately decays and disappears, being resolved into ordinary inorganic compounds. Only as part of a living organism can it be in any sense a basis or supporter of life. Life itself thus remains as an energy, or combination of energies, differing from all others in that while they actuate ordinary matter, it will only actuate organised and protoplasmic matter.

But it may be said, ‘This is after all a familiar thing. We see an egg, or a spore, or a seed made up of a little protoplasm and a few other substances, and it proceeds of itself to grow and shape itself into a complex organism, passing spontaneously through many processes and changes to that end.’ This is true ; but do we ever find such a germ occurring in any other way than as a product of a previous living organism ? We can no more obtain the smallest or simplest egg or spore or the simplest animal or plant directly from dead Nature than we can make a world out of nothing. The previous statements give us some idea of the reason of this. In such a process all would be implied that constitutes the material of the whole of the physical sciences and an unknown quantity beyond, which we can only express as the undiscovered residue of the infinite power and divinity that lie beyond nature. Whether science will ever go so far as to enable us to create living things, or when dead to restore them to life, we cannot tell. That these things are possible, we know, and we may be certain that at some period or periods in the history of the earth living beings originated. We may also be certain that when they originated all the previous arrangements of inorganic nature had been completed and combined to that end ; but what were the details of this we have not at present the means of knowing.

If, then, we find in the little dot of protoplasm that constitutes the egg of an insect the power to develop itself into the parts and structures of the perfect insect, and if one should find that the insect so developed has the further power to modify itself into varietal forms, we may have a vast and interesting field of biological study, but we may still remain ignorant of the origin of the mysterious potencies in the egg, and of the creative processes, extending through untold ages, and of inscrutable complexity and stupendous magnitude, which were necessary to render possible the existence of the egg or the insect.

Such is the problem presented to us by the origin of life ; and it is not too much to say that our modern hypotheses of development, however captivating to the love of simplicity which actuates the general

mind, and however useful as helping to fix the laws
and limits of the variation of living beings, have not
brought us perceptibly nearer to the solution of the
great question, still less to the possibility of solving
it without the power and divinity that lie behind it.
It is of some value, however, to understand the
nature of a question of this kind, even if we cannot
answer it, and we may perhaps best attain to this
kind of information by considering some plain and
simple cases.

Parry, in his Arctic voyages, describes and figures
a remarkable phenomenon witnessed in Greenland
and in other polar and alpine regions, and also to a
modified extent in more temperate climates—that of
the growth of the red snow-plant (*Protococcus
nivalis*).[1]

Large tracts of melting snow on the Greenland
coast are sometimes seen to be coloured with this
humble plant, giving to the previously pure snow a
bright blood-red tint, and often penetrating to some
depth into its mass. Parry informs us that on taking
a bucketful of this snow on board his ship, and
allowing it to melt, the water was seen to contain a
delicate gelatinous matter full of minute grains,
which, under the microscope, resolved themselves
into globular cells with a thin transparent outer wall,
containing a colourless liquid sap, within which was

---

[1] Sometimes referred to genus *Palmella* or to *Chlamydococcus*, and included by Bennett in his family *Protococcaceæ*.

a central protoplasmic mass of a deep red colour, and often divided into still more minute globes, believed to be reproductive germs. Each of these bodies, only one-twelve-thousandth of an inch in diameter, is a perfect plant, capable of performing all the functions of vegetable life and of multiplying in an astonishing manner at a temperature scarcely above the freezing point, and supplied with nourishment and energy by the snow-water and by the solar light and heat. It uses, in short, the power of the solar light and heat to enable it to decompose the small amount of carbonic dioxide and ammonia contained in the melting snow, and to construct from their materials and from water the protoplasm and mucilage and colouring matter necessary to form its own substance. Thus it grows in magnitude, and when mature produces many microscopical germs, which, being discharged from the parent sac, spread themselves on the snow, till from a single germ acres or miles of this may be filled with these tiny organisms.

Here is a low form of plant life existing under what appear to us as unfavourable conditions ; but observe how much it implies. We must have in the first place a pre-existing germ of marvellous potencies, and containing a great number of the complex molecules of protoplasm, and this endowed with life. Next we find this germ possessing chemical powers of a most extraordinary character. The most essential of these, that of decomposing carbon dioxide at a low temperature and with only the help of solar radiation, is thus far impossible to the chemist, and so is also the union of the nascent carbon with other substances to form the mucilage and protoplasm of the sap and the red colouring matter which adorns it. Here is a miracle in the true sense—a mighty work transcending our power and comprehension, and performed by means of an organism the most feeble and apparently inefficient.

If we ask what is the use of this plant, the answer must be—the same with that of the grass of the field. To the few minute animals which can live on melting snow it may serve as food, and washed down into the streams and the sea it helps to sustain the swarming hosts of minute animals of the waters which must have their food provided by the bountiful hand of Nature. But Nature in this sense is only another name for God, whose power and divinity are manifested in every cell of the red snow-plant.

Something, however, may be learnt from the reproduction of this plant. It belongs to a humble group of organisms which must have existed since the dawn of life on our planet, and have continued to propagate themselves throughout the geological ages. Their germs abound in all natural waters and in the air, and are ready to develop themselves whenever the proper conditions can be found. Each set of conditions has also its own special kinds of protophytes fitted for these various conditions, so that there are many genera and species differing in habitat and properties. Even in Greenland, we are informed by Berggren and Dickie, three other species of protophytes are found growing in company with *Protococcus nivalis*, on the ice or the mud and stones upon it. Everywhere these plants form a basis for other and higher kinds of life. When the great eruption of Krakatao had destroyed every living thing, and covered the whole island with barren cinders, the spores of these minute plants, borne to it by the winds and nourished by the rains, developed a coating of vegetation of this kind, on which other and higher plants whose spores and seeds had also been wind- or water-borne immediately developed themselves, presenting an epitome of the first vegetation which clothed our once lifeless continents when the creative fiat, ' Let the earth bring forth plants,' first went forth, but giving no evidence as to the origin of any species of plant *de novo*.

But what of the evolution of the red snow-plant and its congeners? Though there are plants even more simple than the adult red snow-plant, I am not aware that we know any other organism more simple than the microscopic germs or spores of these plants from which they could be derived, and we may as well consider ourselves here face to face with the problem, how can a living cell be produced from inorganic matter, say, from snow-water and the carbonic dioxide and ammonia of the atmosphere, with the aid of solar energy? This problem has been practically solved perhaps many times and under different conditions by creative power, but no evolutionist has yet explained it, and the careful experiments of Pasteur and Tyndall have given only negative results.

These plants are, however, capable of certain variations. The *Protococcus* may differ somewhat in colour, or in proportion of parts in different circumstances, and it is not impossible that some of the forms which have been described as distinct species are really merely varieties of this kind. This might of course enable a botanist to speak of different species of *Protococcus* as having originated by descent with modification ; but if the different forms could be shown to be merely the result of changed conditions, and to be capable of returning in suitable circumstances to the normal properties of the plant, they could not be regarded as true species. He might, however, farther argue that under circumstances of isolation, and where external influences permitted only one form to exist, this might become fixed and continuously reproductive as a distinct species ; but in that case the burden of proof would rest with him, and such proof has not yet been obtained. Until it has, the independent origin of such forms remains quite as possible. If a one-celled *alga* could be produced *de novo* on the surface of Greenland snow, why might not another be independently developed on moist earth or in the water, and why is it necessary to affirm without proof that they have varied from one original?

It is equally impossible to show that these plants have at any time ascended to higher grades. They remain as they were, humble one-celled plants, and may have so continued since the dawn of life on our planet. Even on Krakatao no one supposes that the algoid plants which first took possession changed into higher forms. They only formed a basis on which the spores and seeds of other plants could germinate. Evidently they bring us no nearer to the origin of life, which, as far as they are concerned, is something as primitive and original as that of an atom of oxygen or hydrogen or the force of chemical affinity. Examples of the same kind might be drawn from any of the lower forms of life. None of them give us any mode of transition from the non-living and unorganised to the living and organised, nor do they show any evidence of transition from one grade of organised existence to another.

Something may perhaps be learnt as to the origin of life by a consideration of the probable beginning of some of the organs of animals or plants. I remember when a little boy being suddenly struck on looking at myself in a mirror by the question, How is it that I can see; is not sight a very wonderful thing? I could not answer the question then, and though I have since learnt much as to the laws of light and the physiology of vision, I have not yet fathomed the mysteries of the action of light on nerve-cells and of the transmission of visual impressions to the mind. The eye is indeed one of those wonderful instances of correlation of distinct and distant things which strike us so much in nature. It embodies a vast variety of optical and vital structures and powers, and through the medium of ethereal undulations connects the sentient being with the most distant luminous bodies in the universe.

The eye even in its simplest form is a self-acting and registering instantaneous photographic camera, and having its plates so prepared as to represent colours as well as forms, and it must to this end possess at least a clear refractive medium, photographic pigment-cells, and a nervous apparatus capable of receiving the impressions produced and conveying them to the sensorium. There must have been a time when eyes did not exist. There may have been a time after animals existed when none of them possessed eyes. We have been informed by a leading agnostic evolutionist that we may imagine the eye to have originated spontaneously in some low and simple form, and then ' by the operation of infinite adjustments ' (through infinite time and without any adjustor) to have reached ' the perfection of the eye of the eagle.' Yet this is so little satisfactory that we can well understand the saying of Darwin that the thought of the origin of the eye ' gave him a cold shiver.'

The first appearance of eyes dates very far back in geological time. In the lowest Cambrian rocks, where, for the first time, we find a varied marine fauna, there are crustaceans of the family of trilo-bites with eyes, while there are others in which eyes are not present, or have not been detected. This is parallel with the fact shown in the results of the dredgings of the Challenger, that in the deep sea at present there are some crustaceans furnished with very large eyes, to suit the dim light of the ocean depths, while others living in similar depths are destitute of eyes. Here we have two remarkable facts. First, that in the oldest seas, as well as now, some crustaceans possessed eyes, while others appar-ently living in similar conditions were not so endowed. Secondly, that, so far as known, the eyes of the oldest crustaceans were as complete as those of their modern relatives and on the same plan. With reference to the last statement it is necessary to mention that the eyes of the compound or facetted type which we have in modern crustaceans and insects, and which are of remarkably complex structure, are the oldest known to us. Burmeister long ago showed that the eyes of the ancient trilobites must have possessed all the apparatus found in those of their modern successors, and I have myself seen under the microscope eyes of trilobites of the genus *Phacops* in which the remains of the separate tubes for the several ocelli of the compound eye were plainly discernible. Let it be observed also that the simple or single eye which culminates in the vertebrate animals probably existed as far back as the compound eye, since we have no reason to suppose that the gastropod and cephalopod molluscs which abounded in the Cambrian age were blind, and their eyes must have been of a type distinct in plan from those of the trilobites. The difference between these two kinds of eyes is not in general principle, but in details of plan. In the one a number of small and comparatively simple eyes are grouped together, radiating from a centre, so as to command a wide range of vision without indistinctness in any part. In the other there is but one organ of larger size, and with greater complexity in its apparatus for adjustment to distance and direction. Both these types of eye existed in the Cambrian period with all their essential parts, though perhaps the first mentioned had precedence to some small extent in time. In that early period they were substantially perfected, in so far at least as vision in water is concerned ; and if this perfection arose by 'infinite adjustments,' these must have been made in those pre-Cambrian ages in which we have no evidence of the existence of any creatures requiring to have eyes. Farther, the two types of eyes above referred to must have come in independently. The one could not have originated from the other. It is also to be observed that though the vertebrate eye is on the same general plan with that of the higher molluscs, it differs in some very important details. These vertebrate eyes appeared with the fishes in the Silurian, and I have shown from the structure of an unusually well-preserved eye of a Lower Carboniferous fish (*Palæoniscus*) that in the Palæozoic some of the most minute and delicate arrangements of the eye of the fish already existed.[1] Thus the origin of such organs as the eye becomes as inexplicable on the principle of spontaneous evolution as that of the animal itself.

But while we cannot explain how eyes may be acquired, we know something as to the causes of their being lost, which may perhaps throw light on their origin. A remarkable illustration of this, and also of transmutation as distinguished from origin, and of the equivocal value of the term species as used by evolutionists, is furnished by the cave animals of the great caverns of Kentucky and Virginia, recently so ably described by Packard.[2] These creatures are acknowledged to be merely varietal forms, which by virtue of living for many generations, or it would appear sometimes in a few generations, in the darkness of caverns, have lost the power of vision, and even dispense with eyes, while they have been modified in other respects, as, for example, in the better development of their organs

[1] *Acadian Geology*, Supplement, p. 101.
[2] 'Cave Fauna,' *Memoirs National Academy of Sciences* (U.S.A.), vol. iv.

of touch. No one doubts that they are merely varieties of species living outside the caves, and that, if gradually accustomed to sunlight, they might regain the powers they have lost. They are therefore in no respect truly distinct species, and some of them even pass by imperceptible gradations into the ordinary types from which they arose; yet for convenience of reference distinct specific and even generic names have been given them, and in this way a long list of names of cave fishes, crustaceans, insects, &c. can be made out. These curious creatures cannot therefore be taken as evidence of the origin of new species. [They are no more distinct species of cray-fish, insects, &c. than a blind man is a distinct species of the genus *Homo*. They are clearly merely varietal forms. They cannot be considered to be products of natural selection, but of disuse of certain organs and special demands on others. They have, in short, varied, as Packard explains to us, on the Lamarckian, not on the Darwinian principle. They show the effects of change of conditions of life, and they show great powers of adaptation to new circumstances, acting along with isolation, and the tendency to transmit acquired characters to offspring. Packard even shows reason to believe that they are reproductive with individuals of the ordinary forms of those species which may stray or be carried by floods into these caverns. At the same time many of their peculiarities, as, for instance, the want of colouring matters, are mere physical consequences of the absence of the chemical action of light, and may be induced in the lifetime of an individual, just as a plant may be blanched. Though there is scope for animal life in these caverns, nothing has originated to take advantage of it. Only certain common animals of the daylight, better adapted than others for such conditions, have colonised these recesses, and have undergone certain changes in consequence, which no one can reasonably pretend to be more than varietal. The changes are no more specific than those which certain Arctic animals experience on the approach of winter, and which disappear on the return of spring.

To understand this, let us suppose that at some point in geological time the light of the sun had been gradually extinguished without the entire loss of heat, so that a period of darkness supervened. Under such circumstances many species, both animal and vegetable, might perish. Others, like the cave animals, might survive, and adapt themselves to their new circumstances, becoming colourless, losing their now useless eyes, or portions of them, and improving in delicacy of touch. For generations the whole earth might thus be tenanted by animals like those of the caves. But let us suppose that light was again gradually restored, and that these blind animals recovered the powers and properties they had lost, so that the survivors would present the same appearance as before the period of darkness ; would we have any right to recognise this as the origin of new species ? Would it not rather be a convincing proof of the permanence of specific types ?

Similar evidence to this has been adduced by Darwin himself in the case of pigeons, which after generations of enforced varietal divergence show the capacity to resume even the colouring of the wild original ; and the writer has shown that changes of this kind have passed upon certain marine animals in the Glacial period, and that when this had passed away they resumed their normal characters.[1]

The discussion of the cave animals throws light on the nature of the blind species found in the abysmal depths of the ocean, and also on the strange modifications which befall some common crustaceans when obliged to live in saline waters. It is instructive to note that all these are of the nature of deterioration caused by unfavourable conditions of life.

The same truths apply to the origin of organs in plants. It has been broadly stated by evolutionists that the beauty of flowers is due to the selective action of insects in search of honey. Darwin has said : ' Hence we may conclude that if insects had not been developed on the earth our plants would not have been decked with beautiful flowers, but would have produced only such poor flowers as we see on our fir, oak, nut, and ash trees, on grasses,

[1] *Canadian Record of Science*, January, 1889.

docks, and nettles, which are all fertilised through the agency of the wind.' As Gray has well observed, this at best cannot give us an origin for either flowers or honey-seeking insects. Both must have originated in some different way. All that it can pretend to account for is a certain amount of subsequent change. It fails even to account for this, since the gay flowers are correlated with a vast number of other properties of the plants in question, with which the insects could have nothing to do, and without which they might as well have continued to be fertilised by the wind. Why, indeed, should not the wind be the cause of wind-fertilisation as well as the insects the cause of gay flowers ? And, further, why may not the honey, which in some mysterious way is associated with the gay flowers, be the cause of the suctorial proboscis of the insects, since it surely existed before there were honey-feeding insects, though to a wind-fertilised plant the honey must have been a loss and injury, until it could attract insects by the gay flowers, on the hypothesis, as yet non-existent ? Such hypotheses of natural selection, in short, amount to nothing more than a confusion of correlated natural agencies with causation.

Still another curious question arises with reference to the use of cross-fertilisation. There can be no question that the use of this in nature is not merely to increase the fertility of the individual plants, but so to intermix individual varieties as to keep the

F

species true to its characters. The gardener finds this when he endeavours to select and perpetuate particular varieties. He not only finds that these become less fertile by breeding in and in, so as to tend toward extinction, but that if exposed to the action of the pollen of the normal form, or of another variety, they rapidly return to the type of the species. Thus the processes of wind-fertilisation and insect-fertilisation, which evolution relies on in the interest of descent with modification, are precisely those which the Author of Nature has established to prevent such modification. It would appear that the study of separate organs, whether in plants or animals, as little helps us to any origin, other than that of Divine power, as the study of the organisms as a whole.

It may be said that the result of our inquiry has been eminently unsatisfactory, as failing to show clearly any other origin of species than that ultimate one of the Divine Creative Will. This may be admitted, though what has been said may be held to indicate the path for farther investigation as to the methods of the Creator's action. That evolution is equally powerless in the matter may be shown by the following extract from Darwin :—

Throughout whole classes various structures are formed on the same pattern, and at a very early age the embryos closely resemble each other. Therefore I cannot doubt that the theory of descent with modification embraces all the members of the same great class or kingdom. I believe that animals are descended from at most only four or five progenitors, and plants from an equal or lesser number. Analogy would lead me one step further, namely, to the belief that all animals and plants are descended from some one prototype. But analogy may be a deceitful guide.

Here similarity of plan and early embryonic similarity are taken as evidence of common ancestry. But this is entirely gratuitous, for the first may represent a planning mind following the same ideas in different works ; the second may depend merely on the fact that all ordinary organic development is from the more simple to the more complex. But even if this be granted, the great apostle of evolution still demands four or five primitive species for animals, and about the same for plants. Whence these are to be obtained he cannot tell. Analogy, he says, would lead to one common prototype, but admits that analogy may deceive. It is certain to do so when it proceeds on such data as those he has given ; and even if followed as reliable we have still to ask : Whence, and of what nature, is this prototype, holding within itself the promise and potency of all living things, which are to be unrolled from its almost boundless capacities ?

The questions we have just been considering have led us to think of those ancient animals whose remains are preserved in the rocky strata of the earth, and among which, if anywhere, we should find evidence of the origin of life. I have elsewhere shown that the geological record does not justify us in accepting any of the received theories of descent with modification. The subject is too large for discussion here, but a single illustration from a very familiar animal may show the results to which it leads when we follow the actual guidance of facts, without intercalating, as is the wont of evolutionists, a constant succession of suppositions.

The oyster belongs to the great and ubiquitous class of the bivalve molluscs, and we know representatives of the genus all the way from the Carboniferous to the Modern time, while we know very well the changes of the individual animal from the egg to the perfect form.[1] The oyster begins life as a free-moving creature, without shell, and with those curious movable threads called cilia, by means of which so many humble animalculæ move in the water. In this state it shows little evidence of its future development. When it first assumes a shell, this has already two valves placed on the right and left sides of the animal, but quite different from those of the adult. They are nearly circular, smooth, and marked with regular concentric lines of growth. This is their condition when about a tenth of an inch in diameter. At this stage they resemble the valves of a cockle or a venus shell much more than those of an oyster. Another curious point here is that, while the oyster

[1] Jackson, 'Development of the Oyster,' *Proc. Boston Soc. Nat. Hist.* 1888.

has only one muscle wherewith to close its shell, and many other bivalves have two, the young oyster begins with one, then as it grows a little larger develops two, and later drops one and returns to a single one. The rationale of this is that in the young animal the only muscle needed is the anterior or front one. A little later, as the shell becomes wider, a second, the posterior, is needed. Later still the form and hinge are such that one is sufficient, and the anterior—the one first developed—becomes abortive, while the posterior remains. Other changes might be noticed, but let us think of the significance of these. The egg of the oyster is absolutely undistinguishable from that of any other invertebrate animal. Still it must have within it structures or predetermined powers which denote the animal that is to result from it. The next stage, that of the early embryo, presents a form which we could perhaps decide from its structures to be molluscan ; but we could not tell previous to experience whether, for instance, it would be a uni-valve or a bivalve. The next stages determine it to be a bivalve, but rather one of those with rounded and smooth shells and two muscles than those like the oyster. Here it is to be observed that this dis-tinction of one or two muscles is used to divide the whole of the bivalves into two great groups, so that in this early stage our oyster might be either a *monomyarian* or a *dimyarian*. In this stage it becomes fixed, and begins to spread out its valves into the plaited and unequally valved condition of the adult.

Hence we might make such statements as that the oyster was originally a *monomyarian* with anterior adductor ; but no such mollusc is known in an adult state ; then it was a *dimyarian* with smooth equivalve shell, and of this form are many adult bivalves, both ancient and modern. This is the history of the individuals ; but have we any evidence that it is the history of oysters in geological time ? We know fossil oysters of the ordinary style, though small, as far back as the Carboniferous age, but we know no earlier bivalves having precisely the properties of the early stage. So, though the young spat of these primitive oysters may have been like that of the modern ones, we cannot believe that it came from the eggs of any species known earlier. Still this is possible. Some bivalve of the pre-Carboniferous or Carboniferous age, a *Pterinea* for example, may have produced eggs which, when hatched, attached themselves, and, unlike their parents, produced irregular one-sided shells like the oyster, and their progeny may have continued to do the same. If so, they showed a miraculous persistency in this course of degradation ; and not only so, but in pretty early times, the Jurassic age for example, they had plaited themselves up to an extreme degree of plication and irregularity not surpassed in any subsequent time. Since the Carboniferous, when two so-called species of oysters appear, one in Europe and one in America, so far as we know, these molluscs have not ceased to exist, and at least 200 species are reckoned as known in the fossil state. With respect, then, to these oysters, there may be such suppositions as the following, none of which, however, we can prove.

All these species may have proceeded from one origin, by descent with modification, or the same causes which led to their origination in the Carboniferous may have operated again and again. In like manner the closely allied genera *Exogyra* and *Gryphea*, which existed in the Mesozoic age, may have originated from oysters or may have originated independently. The different so-called species of oysters, which are all very variable, and many of which are scarcely distinguishable, may have had, or some of them may have had, independent origins, or they may be all descendants of the same primeval stock, modifying itself from time to time to suit changed conditions. Thus the oyster is equally to us a miracle, whether it has continued to propagate itself without varying beyond the characters of an oyster through all these vast ages, in which case it is a miracle of heredity; or if from causes to us unknown it has been from time to time developed from animals of some other kind or kinds, in which case it is a miracle of transmutation; or if it has been produced repeatedly without any mediate agency, in which case it is a miracle of creation. It is evident that, while we may imagine any of these possibilities, we cannot establish one more than another, though it is easy, as has been done in the case of the horse and other animals, to forge a chain of derivation by putting together arbitrarily such links as we may select.[1]

In closing this part of our discussion, it is well to observe that we should not be misled in a subject of this kind by vague and general assertions. It is easy to affirm that the lowest animals and the lowest plants are but protoplasm, which is only another name for the chemical compound albumen, and that if we can conceive this to originate from the inorganic union of its elements, we shall have a low form of life from which we can deduce all the higher forms of vital action. In making such affirmation we must take for granted several things, none of which we can yet prove :—(1) That vital force is merely a modification of some of the forces acting on unorganised matter. (2) That such force can be spontaneously originated from other forces without the previous existence of organisation. (3) That, being originated, it has the power to form albumen and other organic compounds. Or, if we prefer another alternative, we may take the following :—(1) That albuminous matter can be produced by the union of its chemical elements without life or organisation. (2) That, being so pro-

[1] See *Story of the Earth and Chain of Life in Geological Times*, by the Author.

duced, it can develop vital forces and organisation, including such phenomena as reproduction, sensation, volition, &c. To believe either of these doctrines in the present state of science is simply an act of faith, not of that kind which is based on testimony or evidence, however slight, but of that unreasoning kind which we usually stigmatise as mere credulity and superstition.

In conclusion, it is a relief to turn from these obscure and uncertain questions to the calm, clear, and decisive statements of revelation already referred to in the first chapter, which, while they give no scientific details and do not in any way hamper the progress of scientific inquiry and discussion, indicate the ultimate conclusions at which this must finally arrive. These we may now further consider in the next chapter, in connection with the origin and development of species in geological time.

CHAPTER IV

THE APPARITION OF SPECIES IN GEOLOGICAL TIME

THE doctrine of organic evolution, whether on the
principle of struggle for existence and natural selection, or on the converse principles of physiological
selection and of progressive adaptation to external
conditions, is essentially biological rather than geological, and has been much more favoured by biologists than by those whose studies lead them more
specially to consider the succession of animals and
plants revealed by the rocks of the earth. These
have for the most part been content to observe the
' apparition,' or first appearance of species, without
inquiry as to their origin or the ultimate causes of
their introduction. Evolutionists, however, require a
great lapse of time for their processes, and thus
come into the discussions of geology. Their demands in this matter have been put in so terse and
clear a manner by a recent advocate that I shall
quote his words as a text or motto for this chapter :
' If art can in a few years effect so great changes in

varietal forms, how much more must Nature be able to effect in the unlimited time at her disposal ? '[1]

In this short sentence we have an epitome of the methods of this interesting philosophy. The writer first assumes what has to be proved, namely, the identity of species and varietal forms. Having thus stolen a march upon us, he next makes the quite unfounded assertion that unlimited time is available for varietal changes. Geologists, no doubt, make large demands on time ; but these are not unlimited. Then we have the human thought and action implied in the word ' art ' placed in comparison with an imaginative personification of Nature, which means nothing unless it is understood to be equal to a personal Creator, who, on the hypothesis, might possibly be dispensed with.

But if the geologist is not convinced by this argument he is asked to consider that in geological time animals and plants have proceeded from more simple to more complex states, and from more generalised forms to those that are more specialised, and that this is in accordance with the analogy of the development from the embryo. He is even accused of stupidity if he fails to be convinced by this analogy, or if he objects that there can be no true analogy between a germ developing from or in a parent, and under special conditions, and an adult animal or series of adult animals supposed

[1] Le Conte, *Evolution*, &c., 1889.

to undergo a similar development under entirely different conditions. It is scarcely too much to say that these preposterous demands are usually made, and tacitly assumed to be granted, in most discussions as to the development of living beings in geological time.

But even if we were to grant these postulates, it would be extremely difficult to fit the actual geological succession into the mould thus arbitrarily prepared for it ; and this we may perhaps be able to illustrate by a few general statements and examples, though its full elucidation would require an extended treatise.[1] We shall, however, be able to see how far this argument falls short of the force of 'demonstration' which has been claimed for it, and shall find some grounds for the doubt with which it has been viewed by many able palæontologists. The complexity of the problems involved has indeed induced many of those most familiar with the succession of life to hold that, while we do not fully know its laws, those that we do comprehend induce the belief that they imply something very different from a continuous and spontaneous evolution. The general truths that we know on this great and complicated subject may be shortly summed up as follows :—

1. Life originated very long ago. If in the Laurentian, or even in the early Cambrian, we can be sure

---

[1] See *Story of the Earth and Chain of Life*, by the Author.

that it is millions of years since the first plant or animal came into existence.

2. The first forms were of low grades, though of high and perfect types within those grades.

3. Many types of animals and plants, perhaps most of the leading groups, have continued without any very manifest change or improvement—have been, in short, fixed or stationary types.

4. Elevation and improvement have taken place by the introduction, apparently in many places simultaneously, of new types, accompanied with, or preceded by, the extinction or degradation of lower forms.

5. Many new forms appear to be introduced at one time and apparently suddenly, so that such groups as the ferns and club-mosses and mares' tails among plants, and at a later date the more perfect fruit-bearing trees, the coral animals, the lamp-shells, the crinoids, the amphibians, the reptiles, the higher mammals enter on the scene abruptly and in large numbers. Thus the impression left on our minds by this grand procession of living beings in geological time is not that of a mere continuous flow, but that of a co-operation of physical agencies toward a particular preparation of our planet, and then the introduction at once and in great force of suitable in-habitants to the abode prepared for them.

This indicates not a mere spontaneous evolution, but a progressive plan carried on by a great variety of causes, some of which we can conjecture, but the greater part of which are still hidden from us, and are only partially and inaccurately presented to us

TABULAR VIEW OF GEOLOGICAL PERIODS AND OF LIFE-EPOCHS.

	Geological Period	Animal Life	Vegetable Life
KAINOZOIC, or NEOZOIC	*Post-Tertiary*	Age of *Man* and *Modern Mammals*	Age of *Angiosperms* and *Palms.*
	Tertiary . .	Age of *Extinct Mammals.* (Earliest Placental Mammals.)	,, ,,
MESOZOIC	*Cretaceous* .	Age of *Reptiles* and *Birds*	(Earliest Modern Trees.) Age of *Cycads* and *Pines.*
	Jurassic . .	,, ,,	,, ,,
	Triassic . .	(Earliest Marsupial Mammals.)	,, ,,
PALÆOZOIC	*Permian* . .	(Earliest True Reptiles.)	Age of *Acrogens* and *Gymnosperms.*
	Carboniferous	Age of *Amphibians* and *Fishes*	,, ,,
	Erian or *Devonian*	,, ,,	,, ,,
	Silurian . .	Age of *Molluscs, Corals,* and *Crustaceans*	(Earliest Land Plants.) Age of *Algæ.*
	Cambrian .	,, ,,	,, ,,
EOZOIC	*Huronian* .	Age of *Protozoa.* (First Animal Remains.)	Indications of Plants not determinable.
	Laurentian .	,, ,,	,, ,,

by any scheme of evolution yet proposed. If in the table above we were to represent diagrammatically the development of animals and plants, this would appear not as a smooth and continuous stream, but as a series of great waves, each rising abruptly, and then descending and flowing on at a lower level along with the remains of those preceding it. This will be explained more in detail in the following pages, in which it may be necessary to mention briefly some of the leading facts ascertained by geology.

Geological investigation has disclosed a great series of stratified rocks composing the crust of the earth, and formed at successive times, chiefly by the agency of water. These can be arranged in chronological order ; and, so arranged, they constitute the physical monuments of the earth's history. We must here take for granted, on the testimony of geology, that the accumulation of this series of deposits has extended over a vast lapse of time, and that the successive formations contain remains of animals and plants, from which we can learn much as to the order of introduction of life on the earth. Without entering into geological details, it may be sufficient to present in the condensed table on the opposite page this grand series of formations, with the general history of life as ascertained from them.

In the oldest rocks known to geologists—those of the Eozoic time—some indications of the presence of life are found. Great beds of limestone are contained in these formations, vast quantities of carbon in the form of graphite, and thick beds of iron-ore. All these are supposed, from their mode of occurrence in later deposits, to be results, direct or indirect, of the agency of life ; and if they afforded no traces of organic forms, still their chemical character would convey a presumption of their organic origin. But additional evidence has been obtained in the presence of certain remarkable laminated forms penetrated by microscopic tubes and canals, and which are supposed to be the remains of the calcareous skeletons of humbly-organised animals akin to the simplest of those now living in the sea. Such animals—little more than masses of living animal jelly—now abound in the waters, and protect themselves by secreting calcareous skeletons, often complex and beautiful and penetrated by pores, through which the soft animal within can send forth minute thread-like extensions of its body, which serve instead of limbs. The Laurentian fossil known as *Eozoon Canadense* may have been the skeleton of such an animal ; and if so, it is the oldest living thing that we know. But if really the skeleton or covering of such an animal, *Eozoon* is larger than any of its successors, and quite as complex as any of them. There is nothing to show that it could have originated from dead matter by any spontaneous action, any more than its modern representatives could do so. There is no evidence of its progress by evolution into any higher form, and the group of animals to which it belongs has continued to inhabit the ocean throughout geological

time without any perceptible advance in rank or complexity of structure. If, then, we admit the animal
nature of this earliest fossil, we can derive from it
no evidence of spontaneous evolution ; and if we
deny its animal nature, we are confronted with a
still graver difficulty in the next succeeding formations.

Between the rocks which contain *Eozoon* and the
next in which we find any abundant remains of life
there is a gap in geological history either destitute of
evidence of life or showing nothing materially in
advance of *Eozoon*. In the Cambrian age, however,
we obtain a vast and varied accession of living things,
which appear at once, as if by a sudden and simul-
taneous production of many kinds of animals. Here
we find evidence that the sea swarmed with living
creatures near akin to those which still inhabit it, and
nearly as varied. Referring merely to leading groups,
we have many species of the soft shellfishes, or crustaceans, and the worms, the ordinary shellfishes, the
sea-stars, and the sponges.[1] In short, had we been
able to drop our dredge into the Cambrian or Silurian
ocean, we should have brought up representatives of
all the leading types of invertebrate life that exist in
the modern seas—different, it is true, in details of
structure from those now existing, but constructed on

---

[1] From the lowest Cambrian beds in which definite and abundant
forms of life are first met with we have all the leading types of marine
invertebrate life, represented by at least 165 species and 67 genera,
according to Walcott.

G

the same principles, and filling the same places in nature.

If we inquire as to the history of this swarming marine life of the early Palæozoic, we find that its several species, after enduring for a longer or a shorter time, one by one became extinct, and were replaced by others belonging to the same groups. Thus there is in each great group a succession of new forms, distinct as species, but not perceptibly elevated in the scale of being. In many cases, indeed, the reverse seems to be the case ; for it is not unusual to find the successive dynasties of life in any one family manifesting degradation rather than elevation. New, and sometimes higher, forms, it is true, appear in the progress of time, but it is impossible, except by violent suppositions, to connect them genetically with any predecessors. The succession throughout the Palæozoic presents the appearance rather of the unchanged persistence of each group under a succession of specific forms, and the introduction from time to time of new groups, as if to replace others which were in process of decay and disappearance.

In the latter half of the Palæozoic we find a number of higher forms breaking upon us with the same apparent suddenness as in the case of the early Cambrian animals. Fishes appear, and soon abound in a great variety of species, representing types of no mean rank, but, singularly enough, belonging, in many cases, to groups now very rare; while the commoner tribes of modern fish do not appear. On the land batrachian reptiles now abound, some of them very high in the sub-class to which they belong. Scorpions, spiders, insects, and millipedes appear, as well as land-snails ; and this not in one locality only, but over the whole northern hemisphere. At the same time the land was clothed with an exuberant vegetation—not of the lowest types nor of the highest, but of intermediate forms, such as those of the pines, the club-mosses, and the ferns, all of which attained in those days to magnitudes and numbers of species unsurpassed, and in some cases unequalled, in the modern world. Nor do they show any signs of an unformed or imperfect state. Their seeds and spores, their fruits and spore-cases, are as elaborately con- structed, the tissues and forms of their stems and leaves as delicate and beautiful, as in any modern plants. Nay, more ; the cryptogamous plants of this age show a complexity and perfection of structure not attained to by their modern successors. So with the compound eyes and filmy wings of insects, the teeth, bones, and scales of batrachians and fishes ; all are as perfectly finished, and many quite as complex and elegant, as in the animals of the present day.

This wonderful Palæozoic age was, however, but a temporary state of the earth. It passed away, and was replaced by the Mesozoic, emphatically the reign of reptiles, when animals of that type attained to

colossal magnitude, to variety of function and structure, to diversity of habitat in sea and on land, altogether unexampled in their degraded descendants of
modern times. Sea-lizards of gigantic size swarmed
everywhere in the waters. On land huge quadrupeds
like Atlantosaurus and Iguanodon and Megalosaurus
greatly exceeded the elephants of later times, and
possessed frames and structures now altogether with-
drawn from the reptiles, and possessed only by
mammals and birds. Some of them walked erect on
their hind feet, others had true horns like the modern
oxen or snout horns like the rhinoceros.[1]   Winged
reptiles—some of them of small size, others with
wings twenty feet in expanse—flitted in the air.
Strangely enough, with these reptilian lords appeared
a few small and lowly mammals, forerunners of the
coming age. Birds also make their appearance, and
at the close of the period forests of broad-leaved trees,
altogether different from those of the Palæozoic age,
and resembling the species of our modern woods,
appear for the first time over great portions of the
northern hemisphere.

The Kainozoic, or Tertiary, is the age of mammals
and of man. In it the great reptilian tyrants of the
Mesozoic disappear, and are replaced on land and sea
by mammals or beasts of the same orders with those
now living, though differing as to genera and species.
So greatly indeed did mammalian life abound in this

[1] *Ceratopsidæ* of Marsh, *Am. Jl. Sci.* 1889.

period that in the middle part of the Tertiary most of the leading groups were represented by more numerous species than at present, while many types then existing have now no representatives. At the close of this great and wonderful procession of living beings comes man himself—the last and crowning triumph of creation, the head, thus far, of life on the earth.

If we imagine this great chain of life, extending over periods of enormous duration in comparison with the short span of human history, presenting to the naturalist hosts of strange forms which he could scarcely have imagined in his dreams, we may understand how exciting have been these discoveries crowded within the lives of two generations of geologists. Further, when we consider that the general course of this great development of life, beginning with Protozoa and ending with man, is from below upward—from the more simple to the more complex —and that there is of necessity in this grand growth of life through the ages a likeness or parallelism to the growth of the individual animal from its more simple to its more complex state, we can understand how naturalists should fancy that here they have been introduced to the workshop of Nature, and that they can discover how one creature may have been developed from another by spontaneous evolution.

We need not be astonished that many naturalists are quite carried away by this analogy, and appear unable to perceive that it is merely a general resem-

blance between processes altogether different in their
nature, and therefore in their causes. The greater
part, however, of the more experienced palæontologists, or students of fossils, have long ago seen that
in the larger field of the earth's history there is very
much that cannot be found in the narrower field of
the development of the individual animal ; and they
have endeavoured to reduce the succession of life to
such general expressions as shall render it more com-
prehensible, and may at length enable us to arrive at
explanations of its complex phenomena. Of these
general expressions or conclusions I may state a few
here, as apposite to our present subject, and as showing how little of real support the facts of the earth's
history give to the pseudo-gnosis of agnostic and
monistic evolution.

1. The chain of life in geological time presents a
wonderful testimony to the reality of a beginning.
Just as we know that any individual animal must
have had its birth, its infancy, and its maturity, and
will reach an end of life, so we trace species and
groups of species to their beginning, watch their culmination, and perhaps follow them to their extinction.
It is true that there is a sense in which geology shows
' no sign of a beginning, no prospect of an end ' ; but
this is manifestly because it has reached only a little
way back toward the beginning of the earth as a
whole, and can see in its present state no indication
of the time or manner of the end. But its revelation

of the fact that nearly all the animals and plants of the present day had a very recent beginning in geological time, and its disclosure of the disappearance of one form of life after another as we go back in time, till we reach the comparatively few forms of life of the Lower Cambrian, and finally have to rest over the solitary grandeur of *Eozoon*, oblige it to say that no living thing known to it is self-existent and eternal.

2. The geological record informs us that the general laws of nature have continued unchanged from the earliest periods to which it relates until the present day. This is the true 'uniformitarianism' of geology, which holds to the dominion of existing causes from the first. But it does not refuse to admit variations in the intensity of these causes from time to time, and cycles of activity and repose, like those that we see on a small scale in the seasons, the occurrence of storms, or the paroxysms of volcanoes. When we find the eyes of the old trilobites to have lenses and tubes similar to those in the eyes of modern crustaceans, we have evidence of the persistence of the laws of light. When we see the structures of Palæozoic leaves identical with those of our modern forests, we know that the arrangements of the soil, the atmosphere, the sunshine, and the rain were the same at that ancient time as at present. Yet, with all this, we also find evidence that long-continued periods of physical quiescence were followed by great

crumplings and foldings of the earth's crust, and we
know that this also is consistent with the operation
of law ; for it often happens that causes long and
quietly operating prepare for changes which may be
regarded as sudden and cataclysmic.

3. Throughout the geological history there is progress toward greater complexity and higher grade,
along with degradation and extinction. Though experience shows that it may be quite possible that
new discoveries may enable us to trace some of the
higher forms of life farther back than we now find
them, yet there can be no question that in the progress of geological time lower types have given place
to higher, less specialised to more specialised. Curiously enough, no evidence proves this more clearly
than that which relates to the degradation of old
forms. When, for example, the reptiles of the Mesozoic age were the lords of creation, there was apparently no place for the larger mammalia which appear
at the close of the reptile dynasty. So in the Palæozoic, when trees of the cryptogamous type predomi-
nated, there seems to have been no room in nature
for the forests of modern type which succeeded them.
Thus the earth at every period was fully peopled
with living beings—at first with low and generalised
structures which attained their maxima at early stages
and then declined, and afterward with higher forms
which took the places of those that were passing
away. These latter, again, though their dominion

was taken from them, were continued in lower positions under the new dynasties. Thus none of the lower types of life introduced was finally abandoned, but, after culminating in the highest forms of which it was capable, each was still continued, though with fewer species and a lower place. Examples of this abound in the history of all the leading groups of animals and plants.

4. There is thus a continued plan and order in the history of life, which cannot be fortuitous, and which is coincident with the gradual perfection of the physical conditions of the earth itself. The chance interaction of organisms and their environment, even if we assume the organisms and environment as given to us, could never produce an orderly continuous progress of the utmost complexity in its detail, and extending through an enormous lapse of time. It has been well said that if a pair of dice were to turn up aces a hundred times in succession, any reasonable spectator would conclude that they were loaded dice ; so if countless millions of atoms and thousands of species, each including within itself most complex arrangements of parts, turn up in geological time in perfectly regular order and a continued gradation of progress, something more than chance must be implied. It is to be observed here that every species of animal or plant, of however low grade, consists of many co-ordinated parts in a condition of the nicest equilibrium. Any change occurring which produces

unequal or disproportionate development, as the experience of breeders of abnormal varieties of animals
and plants abundantly proves, imperils the continued
existence of the species. Changes must, therefore,
in order to be profitable, affect the parts of the
organism simultaneously and symmetrically, and
must be correlated with all the agencies in heaven
and earth that act upon the complex organism and
its several parts. The chances of this may well be
compared to the casting of aces a hundred times in
succession, and are so infinitely small as to be in-
credible under any other supposition than that of
intelligent design.

5. The progress of life in geological time has not
been uniform and uninterrupted. Just as the growth
of trees is promoted or arrested by the vicissitudes of
summer and winter, so in the course of the geological
history there have been periods of pause and acceleration in the work of advancement. This is in accordance with the general analogy of the operations of
nature, and is in no way at variance with the doctrine
of uniformity already referred to. Nor has it anything in common with the unfounded idea, at one
time entertained, of successive periods of entire destruction and restoration of life. Prolific periods of
this kind appear in the marine invertebrates of the
early Cambrian, the plants and fishes of the Devonian,
the batrachians of the Carboniferous, the reptiles of
the Trias, the broad-leaved trees of the Cretaceous,

and the mammals of the early Tertiary. A remarkable contrast is afforded by the later Tertiary and modern time, in which, with the exception of man himself, and perhaps a very few other species, no new forms of life have been introduced, while many old forms have perished. This is somewhat unfortunate, since in such a period of stagnation as that in which we live we can scarcely hope to witness either the creation or the evolution of a new species. Evolutionists themselves—those, at least, who are willing to allow their theory to be at all modified by facts— now perceive this ; and hence we have the doctrine, advanced by Mivart, Le Conte, and others, of ' critical periods,' or periods of rapid evolution alternating with others of greater quiescence. It is further to be observed here that in a limited way and with reference to certain forms of life we can see a reason for these intermittent creations. The greater part of the marine fossils known to us are from rocks now raised up in our continents, and they lived at periods when the continents were submerged. Now, in geological time these periods of submergence alternated with others of elevation ; and it is manifest that each period of continental submergence gave scope for the introduction of numbers of new marine species, while each continental elevation, on the other hand, gave opportunity for the increase of land life. Further, periods when a warm climate prevailed in the Arctic regions—periods when plants such as now live

in temperate regions could enjoy six months of continuous sunshine—were eminently favourable to the
development of such plants, and were utilised for
the introduction of new floras, which subsequently
spread to the southward. Thus we see physical
changes occurring in an orderly succession and made
subservient to the progress of life, and we also see
that, not the adverse conditions of struggle for existence, but the favouring conditions of scope for expansion, were, as might rationally be expected, the
accompaniments and secondary causes of new inbursts
of life.

6. There is no direct evidence that in the course
of geological time one species has been gradually or
suddenly changed into another. Of the latter we
could scarcely expect to find any evidence in fossils ;
but of the former, if it had occurred, we might expect
to find indications in the history of some of the
numerous species which have been traced through
successive geological formations. Species which thus
continue for a great length of time usually present
numerous varietal forms, which have sometimes been
described as new species ; but when carefully scru-
tinised they are found to be merely local and temporary, and to pass into each other. On the other
hand, we constantly find species replaced by others
entirely new, and this without any transition. The
two classes of facts are essentially different, though
often confounded by evolutionists ; and though it is

possible to point out in the newer geological formations some genera and species allied to others which
have preceded them, and to suppose that the later
forms proceeded from the earlier, still, as the connecting links cannot be found, this is mere supposition,
not scientific certainty. Further, it proceeds on the
principle of arbitrary choice of certain forms out of
many, without any evidence of genetic connection.

The worthlessness of such derivation is well shown
in a case which has often been paraded as an illustration of evolution—the supposed genealogy of the
horse. In America a series of horse-like animals has
been selected, beginning with the *Eohippus* of the
Eocene—an animal the size of a fox, and with four
toes in front and three behind—and these have been
marshalled as the ancestors of the fossil horses of
America; for there are no native horses in America
in the modern period, the result of the long series
of improvements having apparently been extinction.
Yet all this is purely arbitrary, and dependent
merely on a succession of genera more and more
closely resembling the modern horse being procurable from successive Tertiary deposits often widely
separated in time and place. In Europe, on the
other hand, the ancestry of the horse has been traced
back to *Palæotherium*—an entirely different form—
by just as likely indications, the truth being that as
the group to which the horse belongs culminated in
the early Tertiary times, the animal has too many

imaginary ancestors. Both genealogies can scarcely
be true, and there is no actual proof of either. The
existing American horses, which are of European
parentage, are, according to the theory, descendants
of *Palæotherium*, not of *Eohippus* ; but if we had not
known this on historical evidence, there would have
been nothing to prevent us from tracing them to the
latter animal. This simple consideration alone is
sufficient to show that such genealogies are not of
the nature of scientific evidence.

This genealogy of the horse has been made so
much of, that perhaps it may be useful to look a
little more minutely into its merits as a 'demonstration' of evolution, and to consider what we really
know of the origin and history of this useful quadruped, so peculiar in some points of structure, and so
eminently the friend and companion of man. It was
immediately preceded in the Tertiary period (Miocene
and Pliocene) by a horse-like animal, the *Hipparion*,
which, among other things, differed from its modern
representative in having its splint bones represented
by two side toes, a conformation supposed to adapt it
to locomotion on soft and swampy ground. The
Hipparion was preceded in the earlier European Tertiary (Eocene) by the *Palæotherium*, and in America
by *Eohippus* and *Orohippus*, in which the side toes
were still further developed so as to touch the ground,
giving the foot a tridactyl character. These relations
have induced the belief that these forms may be an

actual genetic series, the species of *Palæotherium* or *Eohippus* passing through a succession of changes into the modern horse.   Perhaps this case affords as fair an example as we can obtain of the bearing of a derivative hypothesis.   The three genera in question are closely allied.   They succeed each other regularly in geological time.   The horse shows in his splint bones rudiments of organs, which, serving little apparent purpose in him, were more fully developed and of manifest use in his predecessors.   Modern horses have occasionally shown a tendency to develop the side toes, as if returning to the primitive type. Taking this as a fair example of derivation, and admitting, for the sake of argument, its probability, let us consider shortly some of the questions that may be raised with regard to it.   These are principally two :—

1. What limits, if any, must necessarily be set to such an hypothesis, and what relations does it bear to the origin of life at first, and to the succession of animals in geological time ?

2. What causes may be supposed to have led to such derivation ?

Under the first head we have to inquire as to the limits set to derivation by the structure of organic beings themselves, and by the physical conditions and changes which may affect them.   It will be convenient to consider these together.

Supposing that *Palæotherium, Hipparion,* and

Equus, or *Eohippus* and its successors in America are links in a chain extending from the Eocene Tertiary to the present time, can we suppose that by tracing the same series further back it might include any placental mammal? We must answer, Decidedly not, for if the whole time from the Eocene to the present has been required to produce the comparatively small change from *Palæotherium* to horse, the same rate in other cases would carry us back to the Mesozoic period, long before we have any evidence of the existence of 'placental mammals.' In other words, the Tertiary and Modern periods will give us time enough only to effect changes of mammals within the order Ungulata, and perhaps in only one section of that order. The other orders must therefore constitute separate series, and these series must have been advancing abreast of each other. Had each series a separate origin, or is there any mammalian stock in the Mesozoic from which, at the beginning of the Tertiary, these several lines of types may have diverged? Here our information fails. We know only small marsupial and insectivorous mammals in the Mesozoic. On our hypothesis it is possible that these may have been the progenitors of the more varied and advanced marsupials and insectivora of the Tertiary and Modern periods, but scarcely of the placental mammals of the Eocene. There may have been placental mammals, unknown to us, in the Mesozoic, which may constitute the required stock.

The reptiles of the Mesozoic utterly fail to give us the necessary links. If they were changing into anything, it was into birds, not into mammals.

Again, the time in which the horse and its supposed progenitors have lived is one of continuous, unbroken succession of species. More especially in the later Tertiary there seems the best evidence of gradual extinction and introduction of species, without any very widespread and wholesale destruction, and this notwithstanding the intervention of that period of cold and of submergence of land in the northern hemisphere which has given rise to all the much-agitated glacial theories of our time. Can we affirm that such piecemeal work has continued throughout geological time? At this point opens the battle between the catastrophists and uniformitarians in geology, a battle which I am not about to fight over again here. I have elsewhere stated reasons for the belief that neither view can be maintained without the other, and that geological time has consisted of alternations of long periods of physical repose and slow subsidence, in which our more important fossiliferous formations have been deposited, with others of physical disturbance and elevation, with extinction of species. Dana has well shown how completely this view is established by the series of geological formations as seen on the broad area of the American continent. Now the question arises, How would the law of derivation operate in these two

H

different states of our planet ? Let us suppose a state
of things in which far more forms were being destroyed
than were reproduced, another in which introduction
of species was more rapid than extinction. In the
latter case we may suppose an exuberance of new
species to have been produced. In the former there
would be a great clearance of these, and perhaps only
a few types left to begin new series. Do we now
live in one of the periods of diminution or of increase?
Perhaps in the former, since there seems to have been,
in the case of the mammalia of the Post-Pliocene, an
enormous amount of extinction of the grandest forms
of life, apparently without their replacement by new
forms. If so, how far can we judge from our own
time of those which preceded it ? They may have
been far more fertile in new forms, or perhaps farther
in excess in the work of extinction. The question is
further complicated with that which asks if these
differences arise from merely physical agencies acting
on organic beings from without, or if there is in the
organic world itself some grand law of cycles independent of external influences. The answers to such
questions are being slowly and laboriously worked
out by geologists and naturalists, and all the more
slowly that so many inevitable errors occur as to the
specific or varietal value of fossils and the relative
importance of geological facts, while the great gaps
in the monumental history are only little by little
being filled up.

The application of these questions to the animals referred to will serve farther to show their significance as to limitations of derivation. Pictet catalogues eleven species of Eocene *Palæotheria*. Without inquiry as to the origin of these, let us confine ourselves to their progress. Under the hypothesis of derivation, each of these had capacities for improvement, probably all leading to that line of change ending in the production of the horse. If so, then each of our *Palæotheria*, passing through intermediate changes, may be the predecessor of some of the equine animals of the Post-Pliocene and Modern periods. But if, as seems probable, the time intervening between the Eocene and the Modern was unfavourable to the multiplication of such species, then several may have perished utterly in the process, and all might have perished. Supposing, on the contrary, the time to have been favourable to the increase of such creatures, we might have had hundreds of species of equine animals instead of the small number extant at present. Again, what possibilities of change remain in the horse? Can he be supposed capable of going on still farther in the direction of his progress from *Palæotherium* or *Eohippus*, or has he attained a point at which further change is impossible? Will he then, in process of time, wheel round in his orbit, and return to the point from which he set out? Or will he continue unchanged until he becomes extinct? Or can he at a certain point diverge into a new series of

H 2

changes? We do not know any equine animal before
the Eocene. Is it not possible that they may have
originated in some way different from that slow
change by which they are supposed to have been
transmuted into horses, and that in their first origin
they were more plastic than after many changes had
happened to them ? May it not be that the origin of
forms or types is after all something different from
derivative changes, and that new forms are at first plastic, afterwards comparatively fixed—at first fertile in
derivative species, and afterward comparatively barren?
Certainly, unless something of this kind is the case, we
fail to find in the modern world a sufficient number
of representatives of the *Palæotheria, Anoplotheria,
Lophiodons, Coryphodon,* elephants, and mastodons
of the Tertiary. On the other hand, it is scarcely
possible to find a sufficient starting point in the
Eocene for the multitude of cetaceans, carnivores,
ruminants, and quadrumana of the modern time.

The conclusion of this special discussion of the
case of the horse must, I think, be the same as that
arising from our general summary of palæontological
facts, namely, that on the one hand we may not
be justified in affirming that every race of fossil
animals or plants which we may name as a species is
really a distinct product of creation, and that on the
other hand the introduction and extinction of species,
and even of races and varieties, depends on the interaction of causes too numerous and complicated to be

covered by any existing hypotheses of evolution. We may also conclude that the settlement in very early times of so many great principles of construction, and the majestic march of life along determinate paths throughout the vast lapse of geological ages, and along with so many great physical changes, cannot be fortuitous, but must represent a great creative plan conceived in the beginning, and carried out with unchanging consistency.

CHAPTER V

MONISTIC EVOLUTION [1]

WE have already seen that modern evolution in some of its phases is not inconsistent with theism, or even with Christian belief. It is indeed regarded by some of its advocates as a reverent recognition of the mode of development of the plans of Eternal Wisdom, and as capable of throwing light on these plans in the domain of the spiritual as well as in that of the natural. But many of its most ardent advocates, whether scientific or popular, go far beyond the bounds of theism, and enter on atheistic or agnostic speculations, which they regard as the logical and legitimate outcome of the hypothesis of evolution. Perhaps the most eminent advocate of this extreme school is Ernst Haeckel, of Jena, whose views have been presented to the world in his works on *The History of Creation* and *The Evolution of Man*, as well as in many addresses and papers. They may be

---

[1] This chapter has already been published in part in the *Princeton Review* for 1880, p. 444.

taken as the best presentation of monistic, that is, atheistic and materialistic, evolution.

Haeckel is an eminent comparative anatomist and physiologist, who has earned a wide and deserved reputation by his able and laborious studies of the calcareous sponges, the radiolarians, and other low forms of life. In his work on *The Evolution of Man*, he applies this knowledge to the solution of the problem of the origin of humanity, and sets himself not only to illustrate but to 'prove' the descent of our species from the simplest animal types, and even to overwhelm with scorn every other explanation of the appearance of man, except that of spontaneous evolution. The book is full of important facts well stated. The great reputation of the author has given it a wide currency, and it has been translated and reprinted both in England and America, and there can be little doubt that it has exercised an important influence, more especially upon young men of the educated classes, while it has furnished the armoury of many lesser combatants on the same side. It merits, therefore, a careful examination, both as to its data and the manner of treatment of the subject. To understand the latter, it will be necessary in the first place to glance at Haeckel's personal position with reference to the study of Nature.

He is not merely an evolutionist, but what he terms a 'monist,' and the monistic philosophy, as defined by him, includes certain negations and certain positive principles of a most comprehensive and important character. It implies the denial of all spiritual or immaterial existence. Man is to the monist merely a physiological machine, and nature is only a greater self-existing and spontaneously-moving aggregate of forces. Monism can thus altogether dispense with a Creative Will, as originating nature, and adopts the other alternative of self-existence or causelessness for the universe and all its phenomena. Again, the monistic doctrine necessarily implies that man, the animal, the plant, and the mineral are only successive stages of the evolution of the same primordial matter, constituting thus a connected chain of being, all the parts of which sprang spontaneously from each other. Lastly, as the admixture of primitive matter and force would itself be a sort of dualism, Haeckel regards these as ultimately one, and apparently resolves the origin of the universe into the operation of a self-existing energy having in itself the potency of all things. After all, this may be said to be an approximation to the idea of a Creator, but not a living and willing Creator. Monism is thus not identical with pantheism, but is rather a sort of atheistic monotheism, if such a thing is imaginable, and vindicates the assertion attributed to a late lamented physical philosopher, that he had found no atheistic philosophy which had not a God somewhere.

Haeckel's own statement of this aspect of his philosophy is somewhat interesting. He says :—

The opponents of the doctrine of evolution are very fond of branding the monistic philosophy grounded upon it as 'materialism,' by comparing *philosophical* materialism with the wholly different and censurable *moral* materialism. Strictly, however, our own 'monism' might as accurately or as inaccurately be called spiritualism as materialism. The real materialistic philosophy asserts that the phenomena of vital motion, like all other phenomena of motion, are effects or products of matter. The other opposite extreme, spiritualistic philosophy, asserts on the contrary that matter is the product of motive force, and that all material forms are produced by free forces entirely independent of the matter itself. Thus, according to the materialistic conception of the universe, matter precedes motion or active force ; according to the spiritualistic conception of the universe, on the contrary, active force or motion precedes matter. Both views are dualistic, and we hold them both to be equally false. A contrast to both is presented in the *monistic* philosophy, which can as little believe in force without matter as in matter without force.

It is evident that if Haeckel limits himself and his opponents to matter and force as the sole possible explanations of the universe, he may truly say that matter is inconceivable without force, and force inconceivable without matter. But the question arises, What is the monistic power beyond these, the ' Power behind Nature'? and as to the true nature of this the Jena philosopher gives us only vague generalities, though it is quite plain that he cannot admit a spiritual Creator. Further, as to the absence of any spiritual element from the nature of man, he does not leave us in doubt as to what he means ; for, immediately after the above paragraph, he informs us that 'the "spirit" and "mind" of man are but forces which are inseparably connected with the material substance of our bodies.' Just as the motive power of our flesh is involved in the muscular form-element, so is the thinking force of our spirit involved in the form-element of the brain.' In a note appended to the passage he says that monism 'conceives nature as one whole, and nowhere recognises any but mechanical causes.' These assumptions as to man and nature pervade the whole book, and of course greatly simplify the task of the writer, as he does not require to account for the primary origin of nature, or for anything in man except his physical frame, and even this he can regard as a thing altogether mechanical.

It is plain that we might here enter our dissent from Haeckel's method, for he requires us to assume many things which he cannot prove, before we can proceed a single step in the evolution of man. What evidence is there, for example, of the possibility of the development of the rational and moral nature of man from the intelligence and instinct of the lower animals, or of the necessary dependence of the phenomena of mind on the structure of brain-cells? The evidence, as far as it goes, seems to tend the other way. What proof is there of the spontaneous evolution of living forms from inorganic matter? Experiment so far negatives the possibility of this. Even if we give Haeckel, to begin with, a single living cell or granule of protoplasm, we know that this protoplasm must have been produced by the agency of a living vegetable cell previously existing, and we have no proof that it can be produced in any other way. Again, what particle of evidence have we that the atoms or the energy of an incandescent fire-mist have in them anything of the power or potency of life? We must grant the monist all these postulates as pure matters of faith before he can begin his demonstration ; and as none of them are axiomatic truths, it is evident that so far he is simply a believer in the dogmas of a philosophic creed, and weak as other men whom he affects to despise.

We may here place over against his authority that of another eminent physiologist of more philosophic mind, the late Dr. Carpenter, who has said :—

As a physiologist I must fully recognise the fact that the physical force exerted by the body of man is not generated *de novo* by his will, but is derived directly from the oxidation of the constituents of his food. But holding it as equally certain, because the fact is capable of verification by everyone as often as he chooses to make the experiment, that in the performance of every volitional movement physical force is put in action, directed, and controlled by the individual personality or *ego*, I deem it as absurd and illogical to affirm that there is no place for a God in nature, originating, directing, and controlling its forces by His will, as it would be to assert that there is no place in man's body for his conscious mind.

Taking Haeckel on his own ground, as above

defined, we may next inquire as to the method which he employs in working out his argument. This may be referred to three leading modes of treatment, which, as they are somewhat diverse from those ordinarily familiar to logicians, and are extensively used by evolutionists, deserve some illustration, more especially as Haeckel is a master in their use.

An eminent French professor of the art of sleight-of-hand has defined the leading principle of jugglers to be that of 'appearing and disappearing things'; and this is the best definition that occurs to me of one method of reasoning largely used by Haeckel, and of which we need to be on our guard when we find him employing, as he does in almost every page, such phrases as 'it cannot be doubted,' 'we may there-fore assume,' 'we may readily suppose,' 'this after-wards assumes or becomes,' 'we may confidently assert,' 'this developed directly,' and the like, which in his usage are equivalent to the *presto* of the con-juror, and which, while we are looking at one structure or animal, enable him to persuade us that it has been suddenly transformed into something else.

In tracing the genealogy of man he constantly employs this kind of sleight-of-hand in the most adroit manner. He is perhaps describing to us the embyro of a fish or an amphibian, and as we become interested in the curious details, it is suddenly by some clever phrase transformed into a reptile or a bird; and yet without rubbing our eyes and reflecting on the differences and difficulties which he neglects to state, we can scarcely doubt that it is the same animal after all.

The little lancelet, or *Amphioxus*, of the European seas, a creature which was at one time thought to be a sea-snail, but is really more akin to fishes, forms his link of connection between our 'fish ancestors' and the invertebrate animals. So important is it in this respect that our author waxes eloquent in exhorting us to regard it 'with special veneration,' as representing our 'earliest Silurian vertebrate ancestors,' as being of 'our own flesh and blood,' and as better worthy of being an object of 'devoutest reverence' than the 'worthless rabble of so-called "saints."' In describing this animal he takes pains to inform us that it is more different from an ordinary fish than a fish is from a man. Yet as he illustrates its curious and unique structure, before we are aware the lancelet is gone, and a fish is in its place, and this fish with the potency to become a man in due time. Thus a creature intermediate in some respects between fishes and molluscs, or between fishes and worms, but so far apart from either that it seems but to mark the width of the gap between them, becomes an easy stepping-stone from one to the other.

In like manner the ascidians, or sea-squirts, molluscs of low grade, or, as Haeckel prefers to regard them, allied to worms, are most remote in almost every respect from the vertebrates. But in the young state of some of these creatures, and in the adult condition of one animal referred to this group (*Appendicularia*), they have a sort of swimming tail, which is stiffened by a rod of cartilage to enable it to perform its function, and which for a time gives them a certain resemblance to the lancelet or to embryo fishes ; and this usually temporary contrivance, curious as an imitative adaptation, but of no other significance, becomes, by the art of ' appearing and disappearing,' a rudimentary backbone, and enables us at once to recognise in the young ascidian an embryo man.

A second method characteristic of the book, and furnishing indeed the main basis of its argument, is that of considering analogous processes as identical, without regard to the difference of the conditions under which they may be carried on. The great leading use of this argument is in inducing us to regard the development of the individual animal as the precise equivalent of the series of changes by which the species was developed in the course of geological time. These two kinds of development are distinguished by appropriate names. *Ontogenesis* is the embryonic development of the individual animal, and is of course a short process, depending on the production of a germ by a parent animal or parent pair, and the further growth of this germ in connection more or less with the parent or with pro-vision made by it. This is, of course, a fact open to observation and study, though some of its processes are mysterious and yet involved in doubt and uncertainty. *Phylogenesis* is the supposed development of a species in the course of geological time, and by the intervention of long series of species, each in its time distinct, and composed of individuals each going regularly through a genetic circle of its own.

The latter is a process not open to observation within the time at our command ; purely hypothetical, therefore, and of which the possibility remains to be proved, while the causes on which it must depend are necessarily altogether different from those at work in ontogenesis ; and the conditions of a long series of different kinds of animals, each perfect in its kind, are equally dissimilar from those of an animal passing through the regular stages from infancy to maturity. The similarity in some important respects of ontogenesis to phylogenesis was inevitable, provided that animals were to be of different grades of complexity, since the development of the individual must necessarily be from a more simple to a more complex condition. On any hypothesis, the parallelism between embryological facts and the history of animals in geological time affords many interesting and important coincidences. Yet it is perfectly obvious that the causes and conditions of these two successions cannot have been the same. Further, when we consider that the embryo cell which develops into one animal must necessarily be originally distinct in its properties from that which develops into another kind of animal, even though no obvious difference appears to us, we have no ground for supposing that the early stages of all animals are alike ; and when we rigorously compare the development of any animal whatever with the successive appearance of animals of the same or similar groups in geological time, we find many things which do not correspond, not merely in the want of links which we might expect to find, but in the more significant appearance, prematurely or inopportunely, of forms which we would not anticipate. Yet the main argument of Haeckel's book is the quiet assumption that anything found to occur in ontogenetic development must also have occurred in phylogenesis, while manifest difficulties are got rid of by assuming atavisms and abnormalities.

A third characteristic of the method of the book is the use of certain terms in peculiar senses, and as implying certain causes which are taken for granted, though their efficacy and mode of operation are unknown. The chief of the terms so employed are 'heredity' and 'adaptation.' Heredity is usually understood as expressing the power of permanent transmission of characters from parents to offspring, and in this aspect it expresses the constancy of specific forms. But as used by Haeckel it means the transmission by a parent of any exceptional characters which the individual may have accidentally assumed. Adaptation has usually been supposed to mean the fitting of animals for their place in nature, however that came about. As used by Haeckel, it imports the power of the individual animal to adapt itself to changed conditions, and to transmit these changes to its offspring. Thus in this philosophy the rule is made the exception, and the exception the rule, by a skilful use of familiar terms in new senses ; and heredity and adaptation are constantly paraded as if they were two potent divinities employed in constantly changing and improving the face of nature.

It is scarcely too much to say that the conclusions of the book are reached almost solely by the application of the above-mentioned peculiar modes of reasoning to the vast store of facts at command of the author, and that the reader who would test these conclusions by the ordinary methods of judgment must be constantly on his guard. Still, it is not necessary to believe that Haeckel is an intentional deceiver. Such fallacies are those which are especially fitted to mislead enthusiastic specialists, to be identified by them with proved results of science, and to be held in an intolerant and dogmatic spirit.

Having thus noticed Haeckel's assumptions and his methods, we may next shortly consider the manner in which he proceeds to work out the phylogeny of man. Here he pursues a purely physiological method, only occasionally and slightly referring to geological facts. He takes as a first principle the law long ago formulated by Harvey, *Omne vivum ex ovo*—a law which modern research has amply confirmed, showing that every animal, however complex, can be traced back to an egg, which in its simplest state is no more than a single cell, though this cell requires to be fertilised by the addition of the contents of another dissimilar cell, produced either in another organ of the same individual or in a distinct individual. This process of fertilisation Haeckel seems to regard as unnecessary in the lowest forms of life ; but though there are some simple animals in which it has not been recognised, analogy would lead us to believe that in some form it is necessary in all. Haeckel's monistic view, however, requires that in the lowest forms it should be absent, and should have originated spontaneously, though how does not seem to be very clear, as the explanation given of it amounts to little more than the statement that it must have occurred. Still, as a ' dualistic ' process it is very significant with reference to the monistic theory.

Much space is, of course, devoted to the tracing of the special development or ontogenesis of man, and to the illustration of the fact that in the earlier stages of this development the human embryo is scarcely distinguishable from that of lower animals. We may, indeed, affirm that all animals start from cells which, in so far as we can see, are similar to each other, yet which must include potentially the various properties of the animals which spring from them. As we trace them onward in their development, we see these differences manifesting themselves. At first all pass, according to Haeckel, through a stage which he calls the *gastrula*, in which the whole body is represented by a sort of sac, the cavity of which is the stomach, and the walls consist of two layers of cells. It should be stated, however, that many eminent naturalists dissent from this view, and maintain that even in the earliest stages material differences can be observed. In this they are probably right, as even Haeckel has to admit some degree of divergence from this all-embracing 'gastræa' theory. Admitting, however, that such early similarity exists within certain limits, we find as the embryo advances that it speedily begins to indicate whether it is to be a coral animal, a snail, a worm, or a fish. Consequently the physiologist who wishes to trace the resemblances leading to mammals and to man has to lop off, one by one, the several branches which lead in other directions, and to follow that which conducts by the most direct course to the type which he has in view. In this way Haeckel can show that the embryo *Homo sapiens* is in successive stages so like to the young of the fish, the reptile the bird, and the ordinary quadruped that he can produce for comparison figures in which the cursory observer can detect scarcely any difference.

All this has long been known, and has been re-

garded as a wonderful evidence of the homology or
unity of plan which pervades nature, and as constituting man the archetype of the animal kingdom—
the highest realisation of a plan previously sketched
by the Creator in many ruder and humbler forms.
It also teaches that it is not so much in the mere
bodily organism that we are to look for the distinguishing characters of humanity as in the higher rational
and moral nature.

But Haeckel, like other evolutionists of the monistic and agnostic schools, goes far beyond this. The
ontogeny, on the evidence of analogy, as already explained, is nothing less than a miniature representation of the phylogeny. Man must in the long ages
of geological time have arisen from a monad, just as
the individual man has in his life-history arisen from
an embryo-cell, and the several stages through which
the individual passes must be parallel to those in the
history of the race. True, the supposed monad must
have been wanting in all the conditions of origin,
sexual fertilisation, parental influence, and surroundings. There is no perceptible relation of cause and
effect, any more than between the rotation of a car-
riage-wheel and that of the earth on its axis. The
analogy might prompt to inquiries as to common
laws and similarities of operation, but it proves
nothing as to causation.

In default of such proof, Haeckel favours us
with another analogy derived from the science of

language. All the Indo-European languages are believed to be descended from a common ancestral tongue, and this is analogous to the descent of all animals from one primitive species. But unfortunately the languages in question are the expressions of the voice and thought of one and the same species. The individuals using them are known historically to have descended by ordinary generation from a common source, and the connecting links of the various dialects are unbroken. The analogy fails altogether in the case of species succeeding each other in geological time, unless the very thing to be proved is taken for granted in the outset.

The actual proof that a basis exists in nature for the doctrine of evolution founded on these analogies might be threefold. First, there might be changes of the nature of phylogenesis going on under our own observation ; and even a very few of these would be sufficient to give some show of probability. Elaborate attempts have been made to show that variations as existing in the more variable and the domesticated species lead in the direction of such changes ; but the results have been unsatisfactory, and our author scarcely condescends to notice this line of proof. He evidently regards the time over which human history has extended as too short to admit of this kind of demonstration. Secondly, there might be in the existing system of nature such a close connection or continuous chain of species as might at least

strengthen the argument from analogy ; and undoubtedly there are many groups of closely allied
species, or of races confounded with true specific
types, which it might not be unreasonable to suppose
of common origin. These are, however, scattered
widely apart, and the contrary fact of extensive gaps
in the series is so frequent that Haeckel is constantly
under the necessity of supposing that multitudes of
species and even of larger groups have perished, just
where it is most important to his conclusion that they
should have remained. This is of course unfortunate
for the theory, but then, as Haeckel often remarks,
' we must suppose ' that the missing links once existed.
But thirdly, these gaps which now unhappily exist
may be filled up by fossil animals ; and if in the successive geological periods we could trace the actual
phylogeny of even a few groups of living creatures,
we might have the demonstration desired. But here
again the gaps are so frequent and serious that
Haeckel scarcely attempts to use this argument further
than by giving a short and somewhat imperfect summary of the geological succession in the beginning of
his second volume. In this he attempts to give
a series of the ancestors of man as developed in
geological time ; but of twenty-one groups which he
arranges in order from the beginning of the Laurentian to the Modern period, at least ten are not known
at all as fossils, and others do not belong, so far as
known, to the ages to which he assigns them. This

necessity of manufacturing facts does not speak well for the testimony of geology to the supposed phylogeny of man. In point of fact, it cannot be disguised that, though it is possible to pick out some series of animal forms, like the horses already referred to, which simulate a genetic order, the general testimony of palæontology is on the whole adverse to the ordinary theories of evolution, whether applied to the vegetable or to the animal kingdom.[1]

Thus the utmost value which can be attached to Haeckel's argument from analogy would be that it suggests a possibility that the processes which we see carried on in the evolution of the individual may, in the laws which regulate them, be connected in some way more or less close with those creative processes which on the wider field of geological time have been concerned in the production of the multitudinous forms of animal life. That Haeckel's philosophy goes but a very little way toward any understanding of such relations, and that our present information, even within the more limited scope of biological science, is too meagre to permit of safe generalisation, will appear from the consideration of a few facts taken here and there from the multitude employed in these volumes to illustrate the monistic theory.

When we are told that a monad or an embryo-

---

[1] Those who wish to understand the real bearings of palæontology on evolution should study Barraude's *Memoirs on the Silurian Trilobites, Cephalopods, and Brachiopods.*

cell is the early stage of all animals alike, we naturally ask, Is it meant that all these cells are really
similar, or is it only that they appear similar to us,
and may actually be as profoundly unlike as the
animals which they are destined to produce ? To
make this question more plain, let us take the case as
formally stated : ' From the weighty fact that the egg
of the human being, like the egg of all other animals,
is a simple cell, it may be quite certainly inferred
that a one-celled parent form once existed, from
which all the many-celled animals, man included,
ídeveloped.'

Now let us suppose that we have under our
microscope a one-celled animalcule quite as simple in
structure as our supposed ancestor.   Along with this
we may have on the same slide another cell which is
the embryo of a worm, and a third which is the embryo of a man.   All these, according to the hypothesis, are similar in appearance, so that we can by no
means guess which is destined to continue always an
animalcule, or which will become a worm or may
develop into a poet or a philosopher.   Is it meant that
the things are actually alike, or only apparently so ?
If they are really alike, then their destinies must
depend on external circumstances.   Put either of
them into a pond, and it will remain a monad.   Put
either of them into the ovary of a complex animal
and it will develop into the likeness of that animal.
But such similarity is altogether improbable, and it

would destroy the argument of the evolutionist. In this case, he would be hopelessly shut up to the conclusion that 'hens were before eggs;' and Haeckel elsewhere informs us that the exactly opposite view is necessarily that of the monistic evolutionist. Thus, though it may often be convenient to speak of these three kinds of cells as if they were perfectly similar, the method of 'disappearance' has immediately to be resorted to, and they are shown to be in fact quite dissimilar. There is indeed the best ground to suppose that the one-celled animals and embryo-cells referred to have little in common except their general form. We know that the most minute cell must include a sufficient number of molecules of protoplasm to admit of great varieties of possible arrangement, and that these may be connected with most varied possibilities as to the action of forces. Further, the embryo-cell which is produced by a particular kind of animal, and whose development results in the reproduction of a similar animal, must contain potentially the parts and structures which are evolved from it; and fact shows that this may be affirmed of both the embryo and sperm-cells, where there are two sexes. Therefore it is in the highest degree probable that the eggs of a snail and of a man, though possibly alike to our coarse methods of investigation, are as dissimilar as the animals that result from them. If so, the 'egg may be before the hen'; but it is as difficult to imagine the spontaneous production of the egg, which is potentially the hen, as of the hen itself. Thus the similarity of the eggs and early em- bryos of animals of different grades is apparent only ; and this fact, which embodies a great and perhaps in- soluble mystery, invalidates the whole of Haeckel's reasoning on the alleged resemblances of different kinds of animals in their early stages.

A second difficulty arises from the fact that the simple embryo-cell of any of the higher animals rapidly produces various kinds of specialised cells, different in structure and appearance and capable of performing different functions, whereas in the lower forms of life such cells may remain simple, or may merely produce several similar cells little or not at all differentiated. This objection, whenever it occurs, Haeckel endeavours to turn by the assertion that a complex animal is merely an aggregate of inde- pendent cells, each of which is a sort of individual. He thus tries to break up the integrity of the com- plex organism and to reduce it to a mere swarm of monads. He compares the cells of an organism to the ' individuals of a savage community,' who, at first separate and all alike in their habits and occupations, at length organise themselves into a community and assume different avocations. Single cells, he says, at first were alike, and each performed the same simple offices as the others : ' At a later period isolated cells gathered into communities, groups of simple cells, which had arisen from the continued division of a single cell, remained together, and now began gradually to perform different offices of life.'

But this is a mere vague analogy. It does not represent anything actually occurring in nature, except in the case of an embryo produced by some animal which already shows all the tissues which its embryo is destined to reproduce. Thus it establishes no probability of the evolution of complex tissues from simple cells, and leaves altogether unexplained that wonderful process by which the embryo-cell not only divides into many cells, but becomes developed into all the variety of dissimilar tissues evolved from the homogeneous egg, but evolved from it, as we naturally suppose, because of the fact that the egg represents potentially all these tissues as existing previously in the parent organism.

But if we are content to waive these objections, or to accept the solutions given of them by the 'appearance and disappearance' argument, we still find that the phylogeny, unlike the ontogenesis, is full of wide gaps, only to be passed *per saltum*, or to be accounted for by the disappearance of a vast number of connecting links. Of course it is easy to suppose that these intermediate forms have been lost through time and accident ; but why this has happened to some rather than to others cannot be explained. In the phylogeny of man, for example, what a vast hiatus yawns between the ascidian and the lancelet, and another between the lancelet and the lamprey ! It is true that the missing links may have consisted of animals little likely to be preserved as fossils ; but why, if they ever existed, do not some of them remain in the modern seas?   Again, when we have so many species of apes and so many races of men, why can we find no trace, recent or fossil, of that ' missing link ' which we are told must have existed, the ' ape-like men,' known to Haeckel as the ' Alali,' or speechless men ?

A further question which should receive consideration from the monist school is that very serious one : Why, if all is ' mechanical ' in the development and actions of living beings, should there be any progress whatever ?   Ordinary people fail to understand why a world of mere dead matter should not go on to all eternity obeying physical and chemical laws without developing life ; or why, if some low form of life were introduced capable of reproducing simple one-celled organisms, it should not go on doing so.

Further, even if some chance deviations should occur, we fail to perceive why these should go on in a definite manner, producing not only the most complex machines, but many kinds of such machines on different plans, each perfect in its way.   Haeckel is never weary of telling us that to monists organisms are mere machines.   Even his own mental work is merely the grinding of a cerebral machine.   But he seems not to perceive that to such a philosophy the homely argument which Paley derived from the structure of a watch would be fatal. 'The question is whether machines (which monists consider all animals to be, including themselves) infinitely more complicated than watches could come into existence without design somewhere ;'[1] that is, by mere chance. Common-sense is not likely to admit that this is possible.

The difficulties above referred to relate to the introduction of life and of new species on the monistic view. Others might be referred to in connection with the production of new organs. An illustration is afforded, among others, by the discussion of the introduction of the five fingers and toes of man, which appear to descend to us unchanged from the amphibians or batrachians of the Carboniferous period. In' this ancient age of the earth's geological history feet with five toes appear in numerous species of reptilians of various grades. They are preceded by no other vertebrates than fishes, and these have numerous fin-rays instead of toes. There are no properly transitional forms, either fossil or recent, the nearest pectoral fins to fore limbs being those of certain Devonian and Carboniferous fishes; but they fail to show the origin of fingers. How were the five-fingered limbs acquired in this abrupt way? Why were they five rather than any other number? Why, when once introduced, have they continued unchanged up to the

---

[1] Beckett, *Origin of the Laws of Nature.*

present day? Haeckel's answer is a curious example
of his method—

The great significance of the five digits depends on the
fact that this number has been transmitted from the amphibia to all higher vertebrates. It would be impossible to
discover any reason why in the lowest amphibia, as well as
in reptiles and in higher vertebrates up to man, there should
always originally be five digits on each of the anterior and
posterior limbs, if we denied that heredity from a common
five-fingered parent form is the efficient cause of this phenomenon ; heredity can alone account for it. In many amphibia certainly, as well as in many higher vertebrates, we
find less than five digits. But in all these cases it can be
shown that separate digits have retrograded, and have
finally been completely lost. The causes which affected
the development of the five-fingered foot of the higher vertebrates in this amphibian form from the many-fingered
foot (or properly fin) must certainly be found in the adaptation to the totally altered functions which the limbs had to
discharge during the transition from an exclusively aquatic
life to one which was partially terrestrial. While the many-
fingered fins of the fish had previously served almost exclusively to propel the body through the water, they had now
also to afford support to the animal when creeping on the
land. This effected a modification both of the skeleton and
of the muscles of the limbs. The number of fin-rays was
gradually lessened, and was finally reduced to five. These
five remaining rays were, however, developed more vigo-
rously. The soft cartilaginous rays became hard bones ; the
rest of the skeleton also became considerably more firm ;
the movements of the body became not only more vigorous
but also more varied—

and the paragraph proceeds to state other ameliora-
tions of muscular and nervous system supposed to be

related to or caused by the improvement of the limbs.

It will be observed that in the above extract, under the formula 'the causes . . . must certainly be found,' all that other men would regard as demanding proof is quietly assumed, and the animal grows before our eyes from a fish to a reptile as under the wand of a conjuror. Further, the transmission of the five toes is attributed to heredity or unchanged reproduction ; but this, of course, gives no explanation of the original formation of the structure, nor of the causes which prevented heredity from applying to the fishes which became amphibians, and acquired five toes, or to the amphibians which faithfully transmitted their five toes, but not their other characteristics.

It is perhaps scarcely necessary to follow further the criticism of this extraordinary book. It may be necessary, however, to repeat that it contains clear, and in the main accurate, sketches of the embryology of a number of animals, only slightly coloured by the tendency to minimise differences. It may also be necessary to say that in criticising Haeckel we take him on his own ground—that of a monist—and have no special reference to those many phases which the philosophy of evolution assumes in the minds of other naturalists, many of whom accept it only partially or as a form of mediate creation more or less reconcilable with theism. To these more moderate views no reference has been made, though there can be no doubt that many of them are quite as assailable as the position of Haeckel in point of argument. It may also be observed that Haeckel's argument is almost exclusively biological, and confined to the animal kingdom, and to the special line of descent attributed to man. The monistic hypothesis becomes, as already stated, still less tenable when tested by the facts of palæontology. Hence, most of the palæontologists who favour evolution appear to shrink from the extreme position of Haeckel. Gaudry, one of the ablest of this school, in his work on the development of the mammalia, candidly admits the multitude of facts for which derivation will not account, and perceives in the grand succession of animals in time the evidence of a wise and far-reaching creative plan, concluding with the words: 'We may still leave out of the question the processes by which the Author of the world has produced the changes of which palæontology presents the picture.' In like manner the Count de Saporta, in his *World of Plants*, closes his summary of the periods of vegetation with the words :—

But if we ascend from one phenomenon to another, beyond the sphere of contingent and changeable appearance, we find ourselves arrested by a being unchangeable and supreme, the first expression and absolute cause of all existence, in whom diversity unites with unity, an eternal problem insoluble to science, but ever present to the human consciousness. Here we reach the true source of the idea of religion, and there presents itself distinctly to the mind that conception to which we apply instinctively the name of God.

Thus these evolutionists, like many others in America and in England, find a *modus vivendi* between evolution and theism. They have committed themselves to an interpretation of Nature which may prove fanciful and evanescent, and which certainly up to this time remains an hypothesis, ingenious and captivating, but not fortified by the evidence of facts. But in doing so they are not prepared to accept the purely mechanical creed of the monist or to separate themselves from those ideas of morality, of religion, and of sonship to God which have hitherto been the brightest gems in the crown of man as the lord of this lower world. Whether they can maintain this position against the monists, and whether they will be able in the end to retain any practical form of religion along with the doctrine of the derivation of man from the lower animals, remains to be seen Possibly before these questions come to a final issue the philosophy of evolution may itself have been ' modified ' or have given place to some new phase of thought.

In some places there are in Haeckel's book touches of a grim humour which are not without interest, as showing the subjective side of the monistic theory, and illustrating the attitude of its professors to things held sacred by other men. For example, the following is the introduction to the chapter headed

K

' From the Primitive Worm to the Skulled Animal,'
and which has for its motto the lines of Goethe
beginning—

> Not like the gods am I ! full well I know ;
> But like the worms which in the dust must go.

Both in prose and poetry man is very often compared
to a worm ; 'a miserable worm,' 'a poor worm,' are common and also compassionate phrases.   If we cannot detect
any deep phylogenetic reference in this zoological metaphor,
we might at least safely assert that it contains an unconscious
comparison with a low condition of animal development,
which is interesting in its bearing on the pedigree of the
human race.

If Haeckel's reading of Scripture had been sufficiently thorough, he might have quoted here the
melancholy confession of the man of Uz : ' I have
said to the worm, Thou art my mother and my
sister.'   But though Job, like the German professor,
could humbly say to the worm, 'Thou art my mother,'
he could still hold fast his integrity, and believe in
the fatherhood of God.

The moral bearing of monism is further illustrated by the following extract, which refers to a
more advanced step of the evolution—that from the
ape to man, and which shows the honest pride of the
worthy professor in his humble parentage :—

Just as most people prefer to trace their pedigree from
a decayed baron, or if possible from a celebrated prince,
rather than from an unknown humble peasant, so they
prefer seeing the progenitor of the human race in an Adam

degraded by the fall rather than in an ape capable of higher development and progress. It is a matter of taste, and such genealogical preferences do not therefore admit of discussion. It is more to my individual taste to be the more highly developed descendant of an ape, who in the struggle for existence had developed progressively from lower mammals as they from still lower vertebrates, than the degraded descendant of an Adam, God-like but debased by the fall, who was formed from a clod of earth, and of an Eve created from a rib of Adam. As regards the celebrated 'rib,' I must here expressly add, as a supplement to the history of the development of the skeleton, that the number of ribs is the same in man and in woman.[1] In the latter as well as in the former the ribs originate from the skin-fibrous layer, and are to be regarded phylogenetically as lower or ventral vertebræ.[2]

There is no accounting for tastes, yet we may be pardoned for retaining some preference for the first link of the old Jewish genealogical table —'which was the son of Adam, which was the son of God.' As to the 'debasement' of the fall, it is to be feared that the aboriginal ape would object to bearing the blame of existing human iniquities as having arisen from any improvement in his nature and habits ; and it is scarcely fair to speak of Adam as 'formed from a *clod* of earth,' which is not precisely in accordance with the record. As to the ' rib,' which seems so offensive to Haeckel, one would have thought that he would, as an evolutionist, have had some fellow-feeling in this

---

[1] It was scarcely necessary to refer to this childish conception, unless the individual skeleton of Adam had been in question.
[2] Rather, ' vertebral arches.'

K 2

with the writer of Genesis. The origin of sexes is one of the acknowledged difficulties of the hypothesis, and, using his method, we might surely 'assume,' or even 'confidently assert,' the possibility that, in some early stage of the development, the unfinished vertebral arches of the 'skin-fibrous layer' might have produced a new individual by a process of budding or gemmation. Quite as remarkable suppositions are contained in some parts of his own volumes, without any special divine power for rendering them practicable. Further, if only an individual man originated in the first instance, and if he were not provided with a suitable spouse, he might have intermarried with the unimproved anthropoids, and the results of the evolution would have been lost. Such considerations should have weighed with Haeckel in inducing him to speak more respectfully of Adam's rib, especially in view of the fact that in dealing with the hard question of human origin the author of Genesis had not the benefit of the researches of Baer and Haeckel. He had no doubt the advantage of a firm faith in the reality of that Creative Will which the monistic prophets of the nineteenth century have banished from their calculations. Were Haeckel not a monist, he might also be reminded of that grand doctrine of the lordship and superiority of man based on the fact that there was no 'helpmeet for him'; and the foundation of the most sacred bond of human society on the saying of the first man : 'This is now bone of

my bones, and flesh of my flesh.' But monists probably attach little value to such ideas.

It may be proper to add here that, in his references to Adam, Haeckel betrays a weakness, not
unusual with his school, in putting a false gloss on
the old record of Genesis. The statement that man,
was formed from the dust of the ground implies no
more than the production of his body from the common materials employed in the construction of other
animals; this also in contradistinction from the
higher nature derived from the inbreathing or inspiration of God. The precise nature of the method
by which man was made or created is not stated by
the author of Genesis. Further, it would have been
as easy for divine power to create a pair as an individual. If this was not done, and if after the lesson
of superiority taught by the inspection of lower
animals, and the lesson of language taught by naming them, the first man in his ' deep sleep ' is conscious of the removal of a portion of his own flesh, and
then on awaking has the woman ' brought ' to him—
all this is to teach a lesson not to be otherwise learnt.
The Mosaic record is thus perfectly consistent with
itself and with its own doctrine of creation by
Almighty Power.

I have quoted the above passages as examples of
the more jocose vein of the Jena physiologist ; but
they constitute also a serious revelation of the influence of his philosophy on his own mind and heart

in lowering both to a cold, mechanical, and unsym-
pathetic view of man and nature. This is especially
serious when we remember how earnestly, in an
address before the Association of German Naturalists, he advocated the teaching of the methods and
results of this book as those which, in the present
state of knowledge, should supersede the Bible in our
schools. We may well say, with his great opponent
on that occasion, that if such doctrines should be
proved to be true, the teaching of them might become
a necessity, but one that would bring us face to face
with the darkest and most dangerous moral problem
that has ever beset humanity ; and that so long as
they remain unproved it is unwise as well as criminal
to propagate them among the mass of men as conclusions demonstrated by science.

CHAPTER VI

AGNOSTIC EVOLUTION

BETWEEN the position of the materialistic or ener-
gistic monist and that of the theist there are several
stages of so-called agnosticism. The agnostic declines to be called an infidel or an atheist, yet in some
respects he occupies a position more advanced than
either, though expressed in a less offensive way. In
the Christian or New Testament sense an infidel is
merely one who has no faith in Jesus Christ as his
Saviour. He may believe in a God or in many gods.
An atheist may take the farther step of denying the
existence of any god, but may still be open to argue
on the subject. An agnostic may occupy a variety
of positions between that of admitting the possibility
or probability of a First Cause without committing
himself to the doctrine of a personal or living God,
and that of maintaining that it is impossible to have
any knowledge of God, and thereby going beyond
even the standpoint of the atheist. All varieties of
the agnostic creed, or want of creed, necessarily agree
in holding to the spontaneous evolution of the uni-

verse, so that practically agnosticism in some form
and evolution are usually found together.

A recent explanation of Professor Huxley [1]
places agnosticism in the most favourable light in
which it is possible to regard its tenets. He
says :—

> Positively the principle may be expressed : In matters of
> the intellect follow your reason as far as it will take you
> without regard to any other consideration. And nega-
> tively : In matters of the intellect do not pretend that
> conclusions are certain which are not demonstrated or
> demonstrable. That I take to be the agnostic faith,
> which if a man keep whole and undefiled he shall not be
> ashamed to look the universe in the face, whatever the
> future may have in store for him.

To this creed or 'faith,' in so far as intellectual
conclusions are concerned, anyone might subscribe,
but with two reservations, one of them applicable to
each of its clauses. The expression 'follow your
reason' must be taken with the qualification that
there are many cases in which to follow our individual reason against the testimony of those who may
be better informed would be madness ; and the expression 'demonstrated' must be taken with the
qualification that there are in most things different
degrees of probable proof, and that in the majority
of cases we can only adopt the most probable alternative, without insisting on absolute demonstration.

---

[1] *Nineteenth Century*, February, 1889.

It is farther to be observed that this agnostic creed is often held with the mental reservation that nothing must be admitted on any other evidence than that of the senses, and that consequently there are no data for the ascertaining of anything spiritual. This is sometimes put in the offensive form of the statement that science disproves or cannot prove the existence of God, or the spiritual nature and immortality of man. If by this physical science alone is meant, the statement is as foolish as if I were to say that I cannot prove the existence of God by a sum in addition, or the immortality of man by any proposition in the first book of Euclid. Physical science in one aspect of it has nothing to do with such questions. It can, however, supply facts and principles important in ans.vering them.

It is unfortunately this reservation, not explicitly expressed in Professor Huxley's creed, which constitutes the practically important part of the whole matter, and it really amounts to the addition of a third article, to the effect that reason, as informed by natural facts, cannot obtain any demonstration of the existence of God, or of the spiritual nature of man as related to God.

It is my purpose in the following pages to show that physical and natural science perfectly agree with what Christians accept as Divine revelation in establishing the existence and some of the attributes of

God as the living, wise, and almighty Creator, and the loving Father of man.

Herbert Spencer informs us that the 'verbally intelligent' suppositions respecting the origin of the universe are three: (1) It is self-existent. (2) It is self-created. (3) It is created by an external agency. Of the first and second of these suppositions it can scarcely be affirmed that they are even 'verbally intelligent' or conceivable as possible alternatives. That which is self-existent cannot properly be said to have an origin, and an eternal succession of material things is wholly unthinkable. That anything can be self-created seems to be a contradiction in terms. The third supposition is therefore alone tenable, but it is imperfectly expressed, since the cause or agency which produced the universe need not necessarily be 'external,' but may be operative within all its parts as well as without.

If, then, we understand Spencer's third alternative to mean that the 'hypothesis of a First Cause,' to which he elsewhere truly says 'we must commit ourselves,' implies that the universe was created by a power all pervading, and while not limited by the universe still in it as well as without, it comes into exact harmony with the first verse of Genesis, 'In the beginning God created the heavens and the earth.' The writer of that sentence knew that the universe cannot be eternal or self-existent. He knew that it cannot have produced itself. To him,

as to Spencer, the only rational alternative was that it
had been created ; and the name he gives to the Creator,
implying plurality or even infinity in unity, shows
that he regarded this Divine Being as infinite in
power and wisdom, and to him, therefore, known
only in part. It is satisfactory to find that the evolutionist philosopher is shut up by his own method
to the same conclusion with which we have so long
been familiar in the first verse of the Bible, though
he may decline to express it in the same terms, or to
admit the farther teaching of the book.

But from this point the two authorities diverge.
The Bible goes on to give us much information
respecting God and His relations to man. Spencer
stops our way with the dictum that to human reason
the First Cause must be ' wholly inscrutable.' He
thus places us in the dilemma of being obliged to
' commit ourselves ' to the existence of a First Cause,
which must in some way include the potentiality of
all things, but which we cannot even prove to exist.

In this difficulty we may appeal from the agnostic
philosophy, not to revelation, but to natural science.
To science the universe presents phenomena ; but it is
not content to register phenomena. It holds that
' behind every phenomenon there must be a cause,'
and it is of the very essence of science to investigate
these causes. But how can causes be known ? Only
by their effects. We study the phenomena, and from
them we learn the nature and laws of their cause or

causes. If, then, there is a First Cause behind all
material things and energies, it is impossible that we
can be wholly ignorant of the properties of that
cause. We may be sure, not only that it exists, but
that it includes in it potentially all the phenomena
which flow from it. Thus if we know that there is a
cause behind all phenomena, we cannot be agnostics
in reference to that cause. We must be theists, unless we prefer to call ourselves monists or pantheists.

Yet it is not necessary that we should know
everything about the First Cause. Nay, it is impossible
that we should do so unless we first attain to perfect
knowledge of all that it has produced, and this we
know to be impossible. Here, again, revelation is at
one with science. We cannot 'by searching find
out God'; we can know only 'parts of His ways';
He is 'unsearchable.' 'No man hath seen God at
any time.' It is not in that way that we can know
Him, but only in so far as He may have revealed
Himself to us. Yet we have held out to us the grand
and inviting prospect that a time may come when
we shall know even as we are known.

The question remains, How much can we know
of God from nature? In scientific investigations as
to causes, our knowledge of these depends on the
extent of our knowledge of their effects. In the case,
for example, of light and electricity, we have accumu-
lated great stores of observed and experimental facts :
these enable us to arrive at the laws of the energies

which lie behind the facts, and to this extent we can know them ; but we cannot know anything as to their essence, and can only conjecture or calculate their probable effects in circumstances different from those of our observations or experiments. We rise to a higher domain of causation when we investigate the effects of the free will of intelligent beings like men. Human will is no doubt a true and most efficient cause, and it has no doubt ethical laws which regulate its action ; but the difficulties here are greater, and there is perhaps no higher effort of thought than that which relates to the penetration of the plans and counsels of our fellow men, and the principles which actuate them. We rise to a higher plane in the study of God, and need not wonder that here we can know only in part, a mere ' whisper of His ways,' compared with the thunder of His power, as we have it put in that wonderful effort to penetrate the plans of God by the consideration of His dealings with men and things, presented to us by some ancient sage in the Book of Job.

Questions of this kind are not new, though the agnostic philosophy may be a recent phase of human thought ; and it may be interesting to note the way in which the matter is presented to us by a man to whom, as the ' Apostle of the Gentiles,' we owe very much of our modern enlightenment. In that remarkable discussion of the relative degrees of responsibility of the Jew and the heathen, in the early

chapters of the Epistle to the Romans, Paul affirms
with respect to the latter that ' the invisible things of
God, even His eternal power and divinity, can be perceived by them, being understood by the things that
are made.' It will be observed here that the Apostle
refers only to attributes or properties of God as
knowable by us. Of His essence we can know
nothing, any more than we can of the essence of
material things. He also admits that these attributes
are invisible, not objects of sensuous perception, but
only of mental study. He affirms, however, that to
a certain extent they can be known ; and the amount
of the knowledge he expresses by the two terms
power and divinity, the one referring to the energy
manifested in nature, the other to the superhuman
skill and contrivance which it presents to our investigation. This is all that the material universe can
directly teach us of God ; but from this, as he proceeds
to explain, we can inferentially learn more.

This doctrine of Paul has the advantage that it
can appeal to an actual fact in human history, namely,
that men have inferred power and energy as behind
Nature, and that they have usually perceived in its
combinations of means to ends intelligence as well.
If we regard the universe as a mere machine exceed-
ing all our powers of calculation in its magnitude
and gigantic forces, it seems to the last degree absurd
to deny that it presents a manifestation of power.
As the late Dr. Carpenter has well said, an agnostic

is in this respect in the position of a man who, after examining the machinery of a great mill, and finding that the whole is moved by a shaft proceeding from a brick wall, should infer that the shaft is a sufficient cause for what he sees, and that there is no moving power behind it. In like manner, when we consider the variety and intricacy of the parts of the universe, and the manner in which they are co-ordinated to produce certain effects—and this in a way not only beyond our control but beyond our comprehension—we cannot refer this to mere chance, but must admit contrivance and, if so, superhuman skill, and so divinity. This is Paul's contention; and it is so obvious that even agnostics are sometimes inadvertently found to admit it, and are obliged, in spite of themselves, to speak of selection, adaptation, combination, and contrivance in nature. Nature, in short, forces them to speak of divinity in her own language, and not in that of their philosophy.

Farther, since the existence of the universe goes back in time beyond our powers of calculation, we affirm that the power and divinity which it manifests are, as Paul says, 'eternal'; and since this ultimate power can have nothing to determine its action but its own will, we conclude that we are in presence not of brute force, but of what we usually call a personal, but what Paul and the other Bible writers prefer to call a 'living' God. Thus Paul's short statement contains no verbal inaccuracies or inconsistencies, but

leads us at once to what must necessarily be the
kernel of the whole matter, while implying that the
God so well known in some of His attributes cannot
be fully comprehended by us.

A remarkable and curious development of modern
agnosticism, referred to in a previous chapter, is its
attempt to devise some substitute for the religious
beliefs of humanity, which it so inexorably tries to
overthrow and trample on. Two alternatives are
open to it in this direction. One is to make man, as
the head of creation in this world, his own god. This
has been called the religion of humanity. The other
is to turn our attention to the universe as a whole,
and to make it our object of adoration and source of
elevating and ennobling ideas and aspirations. It is
worthy of note that these have been the resources of
mankind in similar circumstances from the earliest
times, for apart from revelation the worship of men
has constantly been given either to deified kings and
heroes or to natural objects, especially to the starry
heavens and the sun. Thus, not knowing the true
God, ancient idolatry and modern agnosticism meet
and worship in the same fane.

It is quite likely that the hero- and star-worship
of primitive humanity was devised by great and good
men of the olden time, relatively as able as our modern
agnostics. Their aim may have been to elevate their
contemporaries and to prepare for a coming age. It
was the grossness and sensuality of the mob that

caused their well-meant efforts to degenerate into stupid superstitions. A similar fate may befall the new religions of our agnostics. But these newer religions have a higher connection. Professor Huxley willingly admits that Jesus Christ is the 'noblest ideal of humanity which mankind has yet worshipped.' If so, why not merge the religion of humanity in the religion of Christ, not in its more debased and degenerate forms, but on the high conception of the New Testament? Spencer believes that we must admit a First Cause, while Huxley, who speaks contemptuously of the religion of humanity, would make the grandeur of the material universe his highest object of adoration. The further admission that this First Cause may be the Almighty Father of mankind would elevate the religion of the universe into theism. Thus it may happen that with larger and more liberal views, even agnosticism may in the future return to the paths of Christian theism, rather than degenerate into a barbarous paganism. The many able men who now profess themselves agnostics have a great and serious responsibility in this matter, for, while many feebler minds may be found to be nearer to the kingdom of heaven than they, others may be enticed into paths where, destitute of Divine guidance, they may be led to darkness and destruction.

CHAPTER VII

THEISTIC EVOLUTION

THIS, in its highest sense, can be nothing less than the development of the divine plan in the construction of the universe ; and as it implies the action of an infinite power behind nature, under the guidance of an omniscient mind, it supplies a full and satisfactory ultimate explanation of phenomena, leaving only for consideration the methods of the development as carried on in time and by such secondary causes as may have been arranged by God.

But such theistic evolution is held in many different ways and in many grades of connection with the Darwinian and other theories. I may select here as one of its latest and ablest exponents Professor Joseph Le Conte, of California, a geologist of some repute and a clear thinker, who aims to combine the various divergent schools of evolution, whether Darwinian or Lamarckian, and to reconcile the whole with theistic beliefs.[1] His proofs of evolution as a law of continuous development of objects and living

[1] *Evolution and its Relation to Religious Thought*, 1889.

beings from one another are not unlike those we have already criticised, and are not so much based on his own science as on the supposed analogies between the development of the individual and the species in biology. We need not deal with these, but may rather notice what is special and peculiar in his view of the matter.

His definition of evolution is somewhat different from that of Spencer and the ordinary Darwinians. Evolution, he says, is (1) continuous progressive change; (2) this is according to certain laws; (3) it is by means of *resident* forces, that is, forces natural to or inherent in the object and its environment. These are, however, forces emanating primarily from a divine power.

Under his first head he unfortunately appears to involve himself in the confusion of the ordinary evolutionists. He states that there are in regard to organic beings three kinds of progressive development. The first is that of the individual from a simple unicellular germ. The second is that implied in the similar gradation from the simplest to the most complex adult animals and plants. The third is that in geological time from the earliest to the modern living beings. It seems here to be taken for granted that all three are similar instances of progressive change of one being into another. Admitting this, of course we at once, as we have already seen, concede all that the evolutionist should fairly be required to prove.

In this Le Conte follows the usual methods of Spencer and Darwin.

Under his second head he notices three laws which he believes to be common to the above kinds of development : (1) The law of differentiation is the same with Spencer's law of progress from the homogeneous to the heterogeneous, and is, though in different ways and degrees, characteristic of development in general. (2) The second law is really a partial exception to the first, and is called the law of progress of the whole, the meaning being that, while on the whole there is progress, a vast number of the lines of development do not rise, but remain stationary or retrograde. This law, be it observed, is one of those which emphasise the difference between the natural development of the embryo and those things with which it is supposed to be analogous ; since, except in rare cases of retrograde development, it does not occur in that of the individual, but it is so frequent in the geological development, as to seem the rule rather than the exception. (3) The third law is applicable only to the third of the great kinds of development, and marks one of its distinctive characters. It is that of rapid culmination and subsequent decadence of the great types of life. A diagram which the author gives to show this is a curious illustration to the eye of the fallacy of the doctrine of gradual and continuous evolution as applied to geological time. It represents, so to speak, successive waves in the development which, as already explained, are very manifest in the geological history, and most instructive, as showing the complex and intermittent progress of organic beings in geological time.[1]

With reference to the third general statement, that the forces causing evolution are resident, the meaning seems to be that they are in some sense natural to or inherent in the being which is in process of modification, or in the objects which environ and act upon it. In one sense—that is, if we include divine action—this is merely asserting the operation of the properties of things, without in any way accounting for them. In another sense, it may be regarded by the monistic and agnostic Darwinian as a surrender of the whole position to their idea of spontaneous and uncaused development. This Le Conte does not intend to do.

In all this we have, though with some important variations, a restatement of the ordinary principles of evolution, and without any adequate analysis of the constituent parts of the diverse supposed kinds of the process. It is scarcely to be wondered at that with these premises Le Conte arrives at the conclusion that evolution is a legitimate induction from the facts of biology, and that it is 'absolutely certain.' We are informed, however, that this absolutely certain evolution is not that of any of the now conflicting schools of thought, but evolution 'as a law of derivation of

[1] Such a view is, of course, very different from the theory of gradual and slow evolution held by Darwinians.

forms from previous forms, as a law of continuity, as a universal law of becoming. In this sense it is not only certain, but axiomatic. It is only necessary to conceive it clearly, to see that it is a necessary truth.' In so far as there is any validity in this statement of the case, it approaches as nearly as theism can to Spencer's hypothesis that all things are self-created. It practically amounts to saying that since, so far as we know, eggs have been produced from birds, and birds from eggs, from time immemorial, it is an axiomatic truth that all things have been thus continuously produced one from another. It is thus evident that Le Conte goes so far, notwithstanding his previous caution, as to place himself at the mercy of the agnostics, who may say that the continuous evolution of things from one another by 'resident force' requires no intervention of a creative power.

Notwithstanding all this Le Conte is a firm believer in God. In his concluding chapters there are some valuable thoughts on the relation of God to nature, and he derives the higher nature of man not from below, but from above. He sees clearly that the forces of nature are ultimately only manifestations of the omnipresent divine energy. He also perceives, what so few seem to comprehend, that this divine energy operates on different planes of being, and limits itself, so to speak, by the prescribed conditions of each, while it can ascend as by a series of steps from its lowest manifestations in dead matter and in the humbler forms of life up to man himself, in whom the image and likeness of God is still limited by his earthly relations and material frame. This great idea of God manifest in nature, but more or less completely in its different grades of being, is the true basis of the doctrine of theistic development. This being understood, the methods by which these different planes of being have been raised one above another and perfected, whether by the complex action of a vast number of co-ordinated secondary causes or by simpler acts of spiritual power, become fair subjects of investigation, whether by science or philosophy, though it is quite likely that they never can be completely understood by finite beings. Man himself occupies merely one plane or grade in the great system, and there may be far higher and more intelligent grades above him. He can hope to know something of the planes that are below him, but not, except by revelation or mere speculation, of those above ; and his comprehension even of those below as compared with that of the Creator Himself must be crude and imperfect.

It further follows that if we regard nature as a manifestation of God, we must not expect to reduce its many lines of progress and advancement to one simple cause or mode. The methods of action of divine power are to our view infinite in variety ; and though we can ascertain their laws and the secondary causes employed, we can know these only in part,

and we know enough to be assured that in the
origin and development of even the humbler
forms of life they may be vastly more multiform and
complex than those employed in the most complicated
combinations of machinery or of process in the works
of man. Newton felt himself to be like a child playing with the sands on the shore of a boundless ocean ;
and in the presence of any organised being we are but
as infants gazing on the mysterious movements of
some intricate machine, and whose thoughts as to its
origin, operation, and uses may be of the crudest
character. Yet one great advantage we have as
theists is that we can hold by the hand of a Father
who knows all the secrets of the mighty fabric which
perplexes us, and can explain to us, little by little, so
much of it as it may be useful to us to know. Thus,
however much we may be mistaken in our first impressions, we may hope to arrive at some measure of
truth, and can find relief from the difficulties of our
own imperfections and of the pressure of our environment, in faith in the loving and all-wise Father of our
spirits.

Le Conte sums up this view of the matter in a
short chapter, which clearly sets forth the compati-
bility of the spiritual world and revelation ; not with
any of the usual theories of evolution, but with
natural law, on the supposition of the divine energy
operating on different planes of being ; and this relation of the natural and spiritual holds equally good

whether we suppose God to have proceeded in His great work by the method of direct development, or, as is more likely, by many methods, more or less diverse :—

If man be indeed something more than a higher species of animal ; if man's spirit be indeed a spark of Divine energy individuated to the point of self-consciousness and recognition of his relation to God ; if spirit-embryo developing in the womb of Nature through all geological time came to birth and independent spirit-life in man, and thus man alone is a *child of God* as well as a product of Nature—if all this be true, then it is evident that this wholly *new* relation requires also a wholly different mode of Divine operation. If God operates on Nature only by regular processes, which we call *natural laws*, then He *must* operate on spirit in a different and a more direct way, and this we call *revelation.* If to the student of nature it is inconceivable that He should operate on nature except by natural laws (for this is the name we give to His chosen mode of operation there), then to the student of theology it is equally inconceivable, if our view of man be true, that He should not operate on spirit in some more direct and higher way, i.e. by revelation.

But some will ask, Is not this a palpable violation of law? I think not. All Divine operations are, must be, according to reason, i.e. according to law. The operation of the divine on the human spirit, i.e. revelation, must therefore be according to law, but a higher law than that which governs Nature, and, therefore, from *the point of view of Nature*, supernatural. There is nothing wholly unique in this. Life is a higher form of force than the physical and chemical. Life phenomena are therefore superphysical, and if we confined the term nature to dead nature, they would be supernatural. So the free, self-determined acts of

spirit on spirit, even of the spirit of man on the spirit of
man, much more the Spirit of God on the spirit of man,
may be according to law, and yet from the natural point of
view be supernatural.   It is true that in the complex phenomena, material and spiritual inextricably woven together,
which go to make up human life Science must ever strive
to reduce as much as possible to material laws, for this is
her domain, and she is bound to extend it ; but, if our
view of man be true, there will always remain a large
residuum of phenomena—a whole world of phenomena—
which will never yield, because clearly beyond her domain.
Standing on the lower material plane, these phenomena are
wholly supermaterial, and therefore incomprehensible from
the material point of view.   We must rise and stand on the
higher plane before these also are reduced to law, but a
higher law than that operating on the lower plane.   If,
therefore, Science insists on banishing the supernatural
from the realm of nature, theology may reasonably insist
on its necessity, *in this sense*, in the realm of morals and
religion.

CHAPTER VIII

GOD IN NATURE

IN discussing the attitude of agnostic evolution, we have seen that its position is rendered untenable by the fact that it has no better evidence of matter and energy in which it believes than of God in whom it declines to believe. Spencer admits that our conception of matter is ' built up or extracted from our experiences of force,' and that it is only by energy that matter ' demonstrates itself as existing.' This second-hand demonstration is, however, perfectly satisfactory to all men, and they never, when of sound mind, refuse to act on their belief. But science must, in considering well its own principles, go much farther than this general creed as to matter. It must believe in different kinds of matter, atoms of different weights, an all-pervading ether, and multitudes of other entities of which it has no better evidence than their observed effects. Science therefore may apply the same reasoning to the human will, to the unseen spiritual world, and to God Himself, if only it can discover effects resulting from their action. It may be profitable to consider here this positive evidence in some of its departments.

The agnostic may say that he is content to regard all nature as a product of law, and that this, being inexorable and unchangeable, excludes the idea of a personal will. A little reflection will show that this position is altogether untenable. The laws of nature are in reality not powers or forces at all, but merely the ways in which energy has been found to act. They are mental generalisations of our own ; and the fact that we are able to form these and to understand nature by their means goes to show the harmony between our mental nature and that of their Author, and so to tell us something of Him. They do not reveal to us the ultimate nature of energy, but merely the mode of its action in whatever way it may have been determined at first.

Nor are such laws necessary. We can imagine them to have been different. They may be different in parts of the universe inaccessible to us. They may even change in process of time. Nor is law at all the reverse of rational will. On the contrary, a world without law or regulated by caprice would be intolerable to rational beings.

Viewed in this way, the theistic conception of law is that it is a voluntary limitation of the power of the Creator in the interest of His creatures. To secure this end, nature must be a perfect machine, all the parts of which are adjusted for permanent and har-

monious action. Nay, rather, it may be compared to
a series of machines, each running independently, like
the trains of a railway, but all regulated and connected
by an invisible guidance, which determines the times
and distance of each, and ordains which shall wait
and give place to others. Even this simile, however, gives us the faintest possible conception of the
countless interactions and interdependencies of natural
laws. Thus the conception of natural law rightly
understood becomes the highest evidence of power
and divinity, and the highest realisation of the plans
of superhuman intelligence.

The notion that when anything has been referred
to natural law the action of God may be dispensed
with in relation to that thing, is merely the survival
of a superstition that God must be capricious and
changeable. On the one hand, while by natural law
God limits His freedom of action in the interest of
the Cosmos and of its intelligent inhabitants, and
while He permits us as rational beings to understand
and utilise in our limited way portions of His plans,
the interactions and adjustments of laws of different
grades are so varied and complex in their scope and
application, and in the combinations of which they
are capable, that it is often impossible for finite minds
to calculate their results, while it is entirely beyond
human power to interfere with their majestic action.
Hence the will, the power, and the divinity of the
Creator and His absolute mastery over His creatures

must ever remain unimpaired by natural law. Further, since we can know so little of law, and have so little power to control the resistless energy embodied in nature, it follows that scope for dependence on God, for miracle and prayer, and for what we in our ignorance call the supernatural, which, though not understood by us, may still be most natural, in the sense of being part of the Divine plan, is practically infinite.

The objection to theism based on natural law may indeed be very well met by Dr. Carpenter's figure of the moving power of the mill referred to in a previous chapter. The man who is content to know that a great shaft passing through a brick wall moves all the machinery, might, if it could be shown that this shaft turns constantly and has always so turned, have some ground for the belief that its motion is spontaneous and uncaused. He might at least assume the position of the agnostic, and say that he was entirely ignorant of any moving power beyond the brick wall. But if it were pointed out to him that the motion of the shaft obeyed certain laws—that it stopped at a certain hour every evening, and renewed work at a certain hour every morning ; that it ceased moving at dinner hour and on Sundays—his agnosticism respecting any power or agency beyond the brick wall would become infinitely more unreasonable ; and this would not be mitigated by the regularity of the several changes or by the possibility of formulating their laws.

Nor can he escape by the magisterial denunciation of theistic ideas as 'anthropomorphic' fancies. All science must in this sense be anthropomorphic, for it consists of what nature appears to us to be, when viewed through the medium of our senses, and of what we think of nature as so presented to us. The only difference is this, that if agnostic evolution is true, science itself only represents a certain stage of the development, and can have no actual or permanent truth ; while, if the theistic view is correct, then the fact that man himself belongs to the unity of nature, and is in harmony with its other parts, gives us some guarantee for the absolute truth of scientific facts and principles.

The idea that nature is a manifestation of mind is so ancient and general that it may almost be considered as an intuition, born spontaneously of our own consciousness of will. It proceeds in any case naturally from the analogy between the operations of nature and those which originate in our own will and contrivance. When men begin to think more accurately, this idea acquires a deeper foundation in the conclusion that nature, in all its varied manifestations, is one vast machine or congeries of machines, too great and complex for us to comprehend, and implying a primary energy infinitely beyond that of man ; and thus the unity of nature points to one Creative Mind.

Even to the savage peoples, in whose minds the

idea of unity has not germinated, or from whose traditions it has been lost, a spiritual essence appears to underlie all natural phenomena, though they may regard this as consisting of a separate spirit or Manitou for every material thing. In all the more cultivated races the ideas of natural religion have taken more definite forms in their theology and philosophy. Dugald Stewart has well expressed the more scientific form of this idea in two short statements :—

' 1. Every effect implies a cause.

2. Every combination of means to an end implies intelligence.'

Unless, then, we are prepared to refer the universe and all its laws and arrangements to mere chance or to absolute necessity, either of which views would be not only irrational, but would involve actual mental confusion, we have no escape from the doctrine of design and the Pauline conclusion that power and divinity are manifested in nature.

It may be profitable to illustrate this great truth under a few definite propositions, and with reference as we proceed to the bearing of these on the various current hypotheses of evolution, but more especially on the evidence of what may be termed ' Mind in Nature ' as an evidence of the power and divinity of its Author.

1. It may be maintained that Nature is an exhibition of regulated and determined power. The first

impression of Nature presented to a mind uninitiated in its mysteries is that it is a mere conflict of opposing forces; but so soon as we study any natural phenonema in detail we see that this is an error, and that everything is balanced in the nicest way by the most subtle interactions of matter and force. We find also that, while forces are mutually convertible and atoms susceptible of vast varieties of arrangement, all this is determined by fixed law, and carried out with invariable regularity and constancy.

The vapour of water, for example, diffused in the atmosphere is condensed by extreme cold and falls to the ground in snowflakes. In these, particles of water, previously kept asunder by heat, are united by cohesive force, and the heat has gone on other missions. But these particles do not merely unite; they geometrise. Like well-drilled soldiers, arranging themselves in ranks, they form themselves, according to regular axes of attraction, in lines diverging at an angle of sixty degrees; and thus the snowflakes are hexagonal plates and six-rayed stars, the latter often growing into very complex shapes, but all based on the law of attraction under the same angles. The frost on the window-panes observes the same law, and so does every crystallisation of water, where it has scope to arrange itself in accordance with its own geometry. But this law of crystallisation gives to snow and ice their mechanical properties, and is connected with a multitude of adjustments of water in a solid state to its place in nature. The same law, varied in a vast number of ways in every distinct substance, builds up crystals of all kinds of minerals and crystalline rocks, and is connected with countless adaptations of different kinds of matter to mechanical and chemical uses in the arts. It is easy to see that all this must have been otherwise, but for the institution of many and complex laws.

A lump of coal at first suggests little to excite interest or imagination ; but the student of its composition and microscopic structure finds that it is an accumulation of vegetable matter representing the action of the solar light on the leaves of trees of the Palæozoic age. It thus calls up images of these perished forests, and of the causes concerned in their production and growth, and in the accumulation and preservation of their buried remains. It further suggests the many ways in which this solar energy, so long sealed up, can be recalled to activity in heat, gas light, steam, and electric light, and how remarkably these things have been related to the wealth and the civilisation of modern nations. I may quote here a graphic passage from a popular paper by Huxley, which admirably draws the picture of provision for man, but unfortunately leaves out the Provider :—

Nature is never in a hurry, and seems to have had always before her eyes the adage, 'Keep a thing long enough, and you will find a use for it.' She has kept her beds of coal for millions of years without being able to find a use

for them ; she has sent them beneath the sea, and the sea-
beasts could make nothing of them ; she has raised them
up into dry land and laid the black veins bare, and still for
ages and ages there was no living thing on the face of the
earth that could see any sort of value in them ; and it was
only the other day, so to speak, that she turned a new
creature out of her workshop, who by degrees acquired
sufficient wits to make a fire, and then to discover that the
black rock would burn.

I suppose that nineteen hundred years ago, when
Julius Cæsar was good enough to deal with Britain as we
have dealt with New Zealand, the primeval Briton, blue
with cold and woad, may have known that the strange black
stone which he found here and there in his wanderings
would burn, and so help to warm his body and cook his
food. Saxon, Dane, and Norman swarmed into the land.
The English people grew into a powerful nation, and Nature
still waited for a return for the capital she had invested in
ancient club-mosses. The eighteenth century arrived, and
with it James Watt. The brain of that man was the spore
out of which was developed the steam engine and all the
prodigious trees and branches of modern industry which
have grown out of this. But coal is as much an essential
of this growth and development as carbonic acid is of a
club-moss. Wanting the coal, we could not have smelted
the iron needed to make our engines, nor have worked our
engines when we got them. But take away the engines, and
the great towns of Yorkshire and Lancashire vanish like a
dream. Manufactures give place to agriculture and pasture,
and not ten men could live where now ten thousand are
amply supported.

Thus all this abundant wealth of money and of vivid
life is Nature's investment in club-mosses and the like so
long ago. But what becomes of the coal which is burnt in
yielding the interest? Heat comes out of it, light comes

M 2

out of it, and if we could gather together all that goes up
the chimney and all that remains in the grate of a thoroughly
burnt coal fire, we should find ourselves in possession of a
quantity of carbonic acid, water, ammonia, and mineral
matters exactly equal in weight to the coal. But these
are the very matters with which Nature supplied the club-
mosses which made coal. She is paid back principal and
interest at the same time ; and she straightway invests the
carbonic acid, the water, and the ammonia in new forms of
life, feeding with them the plants that now live. Thrifty
Nature, surely ! no prodigal, but the most notable of house-
keepers.[1]

All this is true and well told ; but who is Nature,
this goddess, who, since the far-distant Carboniferous
age, has been planning for man ?   Is this not another
name for that Almighty Maker who foresaw and
arranged all things for His people ' before the foundation of the world ' ?   If Huxley did not assure us
that he is an agnostic, we might suspect him from
this passage to be a devout theist, and even an
orthodox Calvinist.

It is plain that ' Nature ' in such a connection represents either a poetical fiction, a superstitious
fancy, or an intelligent creative mind.   It is further
evident that such creative mind must be in harmony
with that of man, though vastly greater in its scope
and grasp in time and space.   This conclusion might
be strengthened by many other examples of the
mute prophecies of past geological periods.

Even the numerical relations observed in nature

[1] *Contemporary Review*, 1871.

teach the same lesson. The leaves of plants are not arranged at random, but in a series of curiously-related spirals, differing in different plants, but always the same in the same species, and regulated by different laws. Similar definiteness regulates the ramification of plants, which depends primarily on the arrangement of the leaves. The angle of ramification of the veins of the leaf is settled for each species of plant ; so are the numbers of parts in the flower and the angular arrangement of these parts.

It is the same in the animal kingdom, such numbers as five, six, eight, ten being selected to determine the parts in particular animals and portions of animals. Once settled, these numbers are wonderfully permanent in geological time. The first known land reptiles appearing in the Carboniferous period have five toes: these appear in the earliest known species in the lowest, as already stated, beds of the Carboniferous. Their predecessors, the fishes, had numerous fin-rays ; but when limbs for locomotion on land were contrived the number five was adopted as the typical one. It still exists in the five toes and fingers of man himself. From these, as is well known, our decimal notation is derived. It did not originate in any special fitness of the number ten, but in the fact that men began to reckon by counting their ten fingers. Thus the decimal system of arithmetic, with all that follows it, was settled millions of years ago, in the Carboniferous period,

either by certain low-browed and unintelligent batrachians or by their Maker.

2. Nature presents to us very remarkable co-
operations of dissimilar and widely separated matters
and forces. I have referred to the numerical arrangements of the leaves of plants ; but the leaf itself, in
its structure and functions, is one of the most remarkable things in nature. Composed of layers of loosely-
placed living cells, with air-spaces between them ;
enclosed above and below with a transparent epidermis, the spaces between the cells communicating
with the atmosphere without, by means of microscopic pores, guarded by cunningly contrived valves,
opening or closing according to the hygrometric state
of the air ; connected with the stem of the plant by a
system of tubes strengthened with spiral fibres or
thickening of their walls within—the structure of the
leaf is, mechanically considered, of extreme beauty
and complexity.

But its living functions are still more wonderful.
Receiving the water from the soil with such materials
as it brings thence in solution, and absorbing carbonic
dioxide and ammonia from the air, the living protoplasm of the leaf-cells has the power of chemically
changing all these substances, and of producing from
them those complicated and otherwise inimitable
organic compounds, of which the tissues of the plant
are built up, and which they also prepare for other
purposes in the plant. The force by which this is

done is that of the solar heat and light, both admitted freely into the interior of the leaf through the transparent epidermis, and therein imprisoned, so as to constitute a powerful storehouse of evaporative and chemical energy. In this way all the materials available for the maintenance of life, whether vegetable or animal, are produced, and no other structure than the living vegetable cell, as it exists in the leaf, has the power to effect these miracles of transmutation.

Here, let it be observed, we have the vegetable cell placed in relation with the system of the plant, with the soil, with the atmosphere and its waters, with the distant sun itself, and the properties of its emitted energies. Let it further be observed that, on the one hand, the chemistry involved in this is of a character altogether different from that which applies to inorganic matter, and, on the other, the products derived from a very few elements embrace all that vast variety of compounds which we observe in plants and animals, and which constitute the material of one of the most complex of sciences, that of organic chemistry. Finally, these complicated structures were produced, and all their relations set up at a very early geological period. In so far as we can judge from their remains and the results effected, the leaves of the Palæozoic period were functionally as perfect as their modern successors.

Of course, the agnostic evolutionist may, if he pleases, attribute all this to fortuitous interactions of the sun, the atmosphere, and the earth, and may provide for what these fail to explain by the assumption of potentialities equivalent to the things produced. But the probability of such an hypothesis becomes infinitely small when we consider the variety and the diversity of things and forces which must have conspired to produce the results observed, and to maintain them so constantly, and yet with so much difference in circumstances and details. It is a relief to turn from such bewildering and gratuitous suppositions to the theory which supposes a designing creative mind.

From the boundless variety of illustrations which the animal kingdom presents I may select one—the contrivances by means of which marine animals are enabled to balance themselves in the waters. In that wonderful hymn of creation, Psalm civ., at whose compass and truth and grandeur the great Humboldt expressed his astonishment, we find in one of the verses mention of the great and wide sea wherein are ' moving things innumerable, small and great animals. There go the ships : there is that leviathan Thou hast made to play therein.' I believe that in this passage the ' ships ' are not those of man, but God's floating things whose home is on the sea. In any case, these floaters are marvellous examples of cunning contrivance. The pearly nautilus is an eminent example.[1]

[1] The uses of the chambers of the nautilus shell have been doubted by some recent observers, but the character of the structures would seem to admit of no other interpretation.

Its coiled shell is divided by partitions into air-chambers so proportioned that the buoyancy of the air is sufficient to counterpoise in sea-water the weight of the animal. There are also contrivances by which the density of the contained air and of the body of the animal can be so modified as slightly to disturb this equilibrium and to enable the creature to rise or sink in the waters. It would be tedious to describe, without adequate illustrations, all the machinery connected with these adjustments. It is sufficient for our purpose to know that they are provided in such a manner that the animal is practically exempted from the operation of the force of gravity. In the modern seas these provisions are enjoyed by only a few species of the genera *Nautilus* and *Spirula*, but in past geological ages far more complex forms existed. Further, this contrivance is very old. We find in the *Orthoceratites* and their allies of the earliest formations these arrangements in their full perfection, and in some forms even more complex than in later types.

The peculiar contrivances observed in the nautilus and its allies are possessed by no other molluscs ; but there is another group of somewhat lower grade, that of the *Ianthinæ*, or violet snails, in which flotation is provided for in a different way. In these animals the shell is perfectly simple, though light, and the floating apparatus consists in a series of horny air-vesicles attached to what is termed the ' foot ' of the

animal, and which are increased in number to suit its
increasing weight as it grows in size. There are
some reasons to believe that this entirely different
contrivance is as old in geological time as the cham-
bered shell of the nautiloid animals.   It was, indeed,
in all probability, more common and adapted to larger
animals in the Silurian period than at present.

Another curious instance not, so far as yet known,
existing at all in the modern world, is that of the remarkable stalked star-fish described by Professor Hall
under the name *Camerocrinus*, and whose remains are
found in the Silurian rocks.   The crinoids, or feather-
stars, are well-known inhabitants of the seas both in
ancient and modern times, but previous to Professor
Hall's discovery they were known only as animals
attached by flexible stems to the sea-bottom or
creeping slowly by means of their radiating arms.
It was not suspected that any of them had committed
themselves to the mercy of the currents suspended
from floats.   It appears, however, that this was actually
realised in the Silurian period, when certain animals
of this group developed a hollow calcareous balloon-
shaped vesicle, from which they could hang suspended
in the water and float freely.   So far as known, this
remarkable contrivance was temporary, and probably
adapted to some peculiarities of the habits and food
of these animals, occurring only in the geological
period in which they existed.

Examples of this sort of adjustment are found in

other types of animal life. In the beautiful Portuguese man-of-war (*Physalia*) and its allies, flotation is provided for by membranous or cartilaginous sacs or vesicles filled with air, and which are the common support of numerous individuals which hang down from them. In some allied creatures the buoyancy required is secured by little sacs filled with oil secreted by the animals themselves, and in ancient zoophytes, known as graptolites, flotation seems to have been effected in some species by air-vesicles supporting a community of animals.

In each of these cases we have a skilful adaptation of means to ends. The float is so constructed as to avail itself of the properties of gases and liquids, and the apparatus is framed on the most scientific principles and in the most artistic manner. That this apparatus is not mechanically put together, and that in each case the instincts and the habits of the animal have been correlated with it, can scarcely be held by the most obtuse intellect to invalidate the evidence of intelligent design.

3. Structures apparently the most simple and often heedlessly spoken of as if they involved no complexity prove, on examination, to be intricate and complex almost beyond conception. In nothing, perhaps, is this better seen than in that much-abused 'protoplasm' which has been made to do duty for God in the origination of life, but which is itself a most laboriously manufactured material. Albumen,

or white of egg—which is otherwise named protoplasm—is a very complicated substance chemically,
and in its molecular arrangements, and when endowed
with life, it presents properties altogether inscrutable.
It is easy to say that the protoplasm of an egg or of
some humble animalcule or microscopic embryo is
little more than a mass of structureless jelly, yet
in the case of the embryo a microscopic dot of this
apparently structureless jelly must contain all the
parts of, say, a bird or a mammal ; but how we may
never know, and certainly cannot yet comprehend.

There are minute animalcules belonging to the
group of flagellate infusoria, some of which, under
ordinary microscopic powers, appear merely as moving
specks, and show their actual structures only under
the highest powers ; yet these animals can be seen to
have an outer skin and an inner mass, to have pul-
sating sacs and reproductive organs, and threadlike
flagella wherewith to swim. Their eggs are, of course,
much smaller than themselves, so much so that some
·of them are probably invisible under the highest
powers employed. Each of them is potentially an
animal, with all its parts represented structurally in
the same way.

Nor need we wonder at this. It has been calcu-
lated that a speck scarcely visible under the most
powerful microscope may contain two million four
hundred thousand molecules of protoplasm. If each
of these molecules were a brick, there would be enough

of them to build a terrace of twenty-five good dwelling houses. But this is supposing them to be all alike ; whereas we know that the molecules of albumen are capable of being of various kinds. Each of these molecules really contains eight hundred and eighty-two atoms, namely four hundred of carbon, three hundred and ten of hydrogen, one hundred and twenty of oxygen, fifty of nitrogen, and two of sulphur and phosphorus.

Now, we know that these atoms may be differently arranged in different molecules, producing considerable difference of properties. Let us try to calculate of how many differences of arrangement the atoms of one molecule of protoplasm are susceptible, and then to calculate of how many changes these different assemblages are capable in a microscopic dot composed of two million four hundred thousand of them. It is scarcely necessary to say that such a calculation, in the multitudes of possibilities involved, transcends human powers of imagination ; yet it raises questions of mechanical and chemical grouping merely, without any reference to the additional mystery of life.

Let it be observed further that this vastly complex material is assumed as if there were nothing remarkable in it by many of the theorists who plausibly explain to us the spontaneous origin of living things. But Nature, in arranging all the parts of a complicated animal beforehand in an apparently structure-

less microscopic *ovum*, has all these vast numbers to
deal with in working out the exact result, and this
not in one case merely, but in multitudes of cases
involving the most varied combinations. We can
scarcely suppose the atoms themselves to have the
power of thus unerringly marshalling themselves to
work out the structures of organisms infinitely varied,
yet all alike after their kinds. If not, then ' Nature '
must be a goddess gifted with superhuman powers of
calculation and marvellous deftness in arranging invisible atoms.

4. The beauty of form, proportion, and colouring
that abounds in nature affords evidence of mind.
Herculean efforts have been made by modern
agnostic evolutionists to eliminate altogether the
idea of beauty from nature by theories of sexual
selection and the like, and to persuade us that beauty
is merely utility in disguise, and even then only an
accidental coincidence between our perceptions and
certain external things. But in no part of their
argument have they more signally failed in accounting for the observed facts, and in no part have they
more seriously outraged the common-sense and
natural taste of men. In point of fact, we have here
one of those great correlations belonging to the unity
of nature—that indissoluble connection which has
been established between the senses and the æsthetic
sentiments of man and certain things in the external
world. But there is more in beauty than this merely

anthropological relation. Certain forms, for example, adopted in the skeletons of the lower animals are necessarily beautiful because of their geometrical proportions. Certain styles of colouring are necessarily beautiful because of harmonics and contrasts which depend on the essential properties of the waves of light. Beauty is thus, in a great measure, independent of the taste of the spectator. It is also independent of mere utility, since, even if we admit that all these combinations of forms, motions, and colours which we call beautiful are also useful, it is easy to perceive that the end could often be attained without beauty.

It is a curious fact that some of the simplest animals—as, for example, sponges and foraminifera— are furnished with most beautiful skeletons. Nothing can exceed the beauty of form and proportion in the shells of some foraminifera and polycistina, or in the skeletons of some silicious sponges, while it is obvious that these humble creatures, without brains and external senses, can neither contrive nor appreciate the beauty with which they are clothed.

Here I may pause to remark that no feature of the current evolution seems more objectionable than that which refers beauty to low forms of utility, and to selection exercised by animals which can have no intelligent knowledge even of that which attracts them. To an insect a bright spot of any kind would have been as effectual a mark of a honey-bearing flower as the most delicate and elaborate pencilling of colour. To attribute the marvellous beauty of an Argus pheasant or of a bird of Paradise to the taste of the hen bird can scarcely be characterised as anything short of a base superstition ; while it is absolutely irrational as a matter of science, since it is attributing an effect to that which cannot be an efficient cause.

Most persons have seen the beautiful *Euplectella aspergillum*, or ' Venus' flower-basket,' now somewhat common in museums and private collections, but few perhaps have minutely examined its structure. A little observation enables us to see its regular cylindrical form and graceful cornucopia-like curves, combining strength with beauty ; its framework of delicate silicious threads, some regularly placed in vertical bundles, others crossing them, so as to form rectangular meshes, and still others placed diagonally, so as to convert the square meshes into a lace-like pattern. Without this framework are accessory spicules placed in spiral frills, and at the top is a singular network of silicious fibres closing the aperture, while there are long silky threads forming roots below. This structure, so marvellous in the mechanical and æsthetic principles embodied in it, is the skeleton of a sponge ; a soft, slimy, almost structureless creature, which we find it difficult to believe in as a veritable animal ; yet it is the law of this creature, developed from a little oval or sac-like germ, destitute of all trace of the subsequent structures, to produce this wonderful framework. Can anyone who studies such an organism summon faith enough in atoms and forces to believe that their insensate action is the sole cause of its being? But our *Euplectella aspergillum* is only one of several species, and there are other genera more or less resembling it, most of them inhabiting the depths of the sea. All of these build up silicious skeletons on what is termed the hexactinellid plan, but with differences of detail perfectly constant in each species, though we cannot trace these differences to anything corresponding in the animals, nor can we assign them to any property of silica, since the material of the spicules is in a colloidal or uncrystalline state, and the forms are quite different from the crystalline forms of silica.

These hexactinellid sponges have a history. They are widely diffused in our present seas. The chalk formation of Europe abounds with them, and presents forms even more varied and beautiful than those now existing, but which must have lived at a time when large parts of our present continents were in the depths of the ocean. Still further back, in the Silurian age, they seem to have been nearly equally abundant. I have recently studied the microscopic structures of a large collection from the Niagara limestone, consisting of many species, each of which presents arrangements of spicules as beautiful and complex as those of the modern kinds. Still farther back, in the rocks of the Levis division of the Siluro-Cambrian system in Canada, I have found in a single thin bed of shale, representing a muddy sea-bottom of that age, a dozen species of several genera, all bearing testimony to the perfection of this plan of structure at that early date. Salter and Matthew have found in still older Cambrian rocks species of these sponges having delicate spicules still retaining their arrangement, and showing that this beautiful contrivance for the support of a gelatinous animal existed in all its perfection almost at the dawn of life. Through all these vast periods of geological time the hexactinellids have continued side by side with the lithistid sponges, their allies ; and contemporaneously with them the rhizopods and radiolarians, still more simple forms, have built up other styles of skeletons equally wonderful and inexplicable, and embodying other mechanical plans and other types of beauty.

It is scarcely too much to say that no sane mind having presented to it, not as above merely in a few words, but in the actual facts as they might be illustrated with specimens and figures, all this unity and variety, mechanical contrivance and varied beauty, associated with so little of vitality and complexity in the animals concerned, could doubt for a moment the action of a creative intelligence in the initiation of such phenomena, or could believe that they have resulted from the fortuitous interaction of atoms.

Still, admitting this, we are not prevented from attributing something to environment and to reproductive continuity. The waters ' brought forth ' these animals of old, and it is true we cannot conceive of creatures so constructed as living out of the waters. The sea also furnished to them the material out of which to construct their skeletons, either directly or through the medium of still simpler organisms. A.l this and much more respecting the surrounding medium science can understand, though it does not thereby learn the origin of these forms or the reason of their complexity and variety. These do not depend on the properties either of the waters or the silica.

Further, our sponge has the power of increasing and multiplying to replenish the waters. It begets new organisms in its own likeness, and with all its own wonderful powers of unconscious construction. Nay, more, we can see that in this continuous reproduction it has a certain versatility, enabling it to conform to circumstances, and so to present individual and race characters within the species. May not, then, the creative act have been limited to the production of the first hexactinellid, and may not the others have originated by ordinary generation? Here we may admit that, for aught that we know, not only varietal forms, but even some of those which, as met with in successive geological formations, we regard as species may have had a common origin in this way ; but we have no right to affirm this till we have proved it, and we have no right even then to affirm it of other and distinct lines of being, which have gone on parallel with our hexactinellids for indefinite times, and which the very fact of the persistence of the latter within their own cycle of characters would tend to refer to independent origins.

Such, in short, would be the bearing, not of metaphysical arguments, but of the testimony of facts as presented by the structures and history of any group of the lower animals.

5. The instincts of the lower animals imply a higher intelligence.   Instinct, on the theistic view of nature, can be nothing less than a divine inspiration, placing the animal in relations to other things and processes of the most complex character, and which it could not have designed itself.   Further, instinct is by its very nature a thing unimprovable.   Like the laws of nature, it operates invariably, and if diminished or changed it would prove useless for its purpose.   It is not like human inventions, slowly perfected under the influence of thought and imagination, and laboriously taught by each generation to its successors.   It is inherited by each generation in all its perfection, and from the first goes directly to its end as if it were a merely physical cause.

The favourite explanation of instinct from the side of agnostic evolution is that it originated in the struggle for existence of some previous generation,

and was then perpetuated as an inheritance. But,
like most of the other explanations of this school, this
quietly takes for granted what should be proved.
That instinct is hereditary is true ; but the question
is how it began, and to say simply that it did begin
at some time is to tell us nothing. From a scientific
point of view the invariable operation of any natural
law affords no evidence of any gradual or sudden
origin of it at some point of past time ; and when
such law is connected with a complex organism and
various other laws and processes of the external
world, the supposition of its slowly arising from
nothing through many generations of animals becomes absurd in its inefficiency and complexity.
Instinct must have originated in a perfect condition,
and with the organism and its environment already
established. A consideration of any of the almost
countless modifications of instinct in the lower
animals would show this. I shall borrow a very
apposite one from the remarkable work of the Duke
of Argyll on the *Unity of Nature*, which deserves
careful study by everyone who values common-sense
views on the subject :—

On a secluded lake in one of the Hebrides I observed
a dun-diver, or female of the red-breasted merganser (*Mer-
gus serrator*), with her brood of young ducklings. On giving
chase in the boat, we soon found that the young, although
not above a fortnight old, had such extraordinary powers of
swimming and diving that it was almost impossible to cap-
ture them. The distance they went under water and the

unexpected places from which they emerged, baffled all our
efforts for a considerable time. At last one of the brood
made for the shore, with the object of hiding among the
grass and heather which fringed the lake. We pursued it
as closely as we could ; but when the little bird gained the
shore our boat was still about twenty yards off. Long
drought had left a broad margin of small flat stones and
mud between the water and the usual bank. I saw the little
bird run up a couple of yards from the water and then suddenly disappear. Knowing what was likely to be enacted,
I kept my eye fixed on the spot, and when the boat was run
upon the beach I proceeded to find and pick up the chick.
But on reaching the place of disappearance no sign of the
young merganser was to be seen. The closest scrutiny, with
the certain knowledge that it was there, failed to enable me
to detect it. Proceeding cautiously forward, I soon became
convinced that I had already overshot the mark, and on
turning round it was only to see the bird rise like an apparition from the stones, and, dashing past the stranded boat,
regain the lake, where, having now recovered its wind, it
instantly dived and disappeared. The tactical skill of the
whole of this manœuvre and the success with which it was
executed were greeted with loud cheers from the party ; and
our admiration was not diminished when we remembered
that some two weeks before that time the little performer
had been coiled up inside the shell of an egg, and that
about a month before it was apparently nothing but a mass
of albumen and of fatty oils.

On this the Duke very properly remarks that all
idea of training and experience is absolutely ex-
cluded, because it 'assumes the pre-existence of the
very powers for which it professes to account'! He
then turns to the idea that animals are *automata*, or
'machines.' Here it is to be observed that the essen-

tial conception of a machine is twofold. First, it is a merely mechanical structure, put together to do certain things; secondly, it must be related to a combiner and constructor. If we think proper to call the young merganser a machine, we cannot admit the first of these characters without also admitting the second—more especially as the bird is in every way a more complex and marvellous machine than any of human contrivance. He concludes his notice of this attempt at explanation in the following suggestive words :—

Passing now from explanations which explain nothing, is there any light in the theory that animals are '*automata*'? Was my little dipper a diving machine? It seems to me that there is at least a glimmer shining through this idea—a glimmer as of a real light struggling through a thick fog. The fog arises out of the mists of language—the confounding and confusion of meanings literal with meanings metaphorical, the mistaking of partial for complete analogies. 'Machine' is the word by which we designate those combinations of mechanical force which are contrived and put together by man to do certain things. One essential characteristic of them is that they belong to the world of the not living; they are destitute of that which we know as life, and of all the attributes by which it is distinguished Machines have no sensibility. When we say of anything that it has been done by a machine, we mean that it has been done by something which is not alive. In this literal signification it is therefore pure nonsense to say that any living being is a machine. It is simply a misapplication of language, to the extent of calling one thing by the name of another thing, and that other so different as to be its oppo-

site or contradictory. There can be no reasoning, no clear-
ing up of truth, unless we keep definite words for definite
ideas. Or if the idea to which a given word has been ap-
propriated be a complex idea, and we desire to deal with one
element only of the meaning separated from the rest, then
indeed we may continue to use the word for this selected
portion of its meaning, provided always that we bear in
mind what it is that we are doing. This may be, and often
is, a necessary operation, for language is not rich enough to
furnish separate words for all the complex elements which
enter into ideas apparently very simple ; and so of this word
machine. There is an element in its meaning which is
always very important, which in common language is
often predominant, and which we may legitimately choose
to make exclusive of every other. This essential element in
our idea of a machine is that its powers, whatever they may
be, are derived, and not original. There may be great
knowledge in the work done by a machine, but the knowledge is not in it ; there may be great skill, but the skill is
not in it ; great foresight, but foresight is not in it ; in short,
great exhibition of all the powers of mind, but the mind is
not in the machine itself. Whatever it does is done in
virtue of its construction, which construction is due to a
mind which has designed it for the exhibition of certain
powers and the performance of certain functions. These
may be very simple, or they may be very complicated ; but
whether simple or complicated the whole play of its operations is limited and measured by the intentions of its constructor. If he be himself limited, either in opportunity or
knowledge or in power, there will be a corresponding limitation in the things he invents and makes. Accordingly, in
regard to man, he cannot make a machine which has any of
the gifts or powers of life. He can construct nothing which
has sensibility or consciousness, or any other of even the
lowest attributes of living creatures. And this absolute de-
stitution of even apparent originality in a machine, this entire

absence of any share of consciousness, or of sensibility, or of will, is one part of our very conception of it. But that other part of our conception of a machine which consists in its relation to a contriver and constructor is equally essential, and may, if we choose, be separated from the rest, and may be taken as a representative of the whole. If, then, by any agency in nature, or outside of it, which can contrive and build up structures endowed with the gifts of life ; structures which shall not only digest but which shall also feel and see ; which shall be sensible of enjoyment from things condu- cive to their welfare, and of alarm on account of things which are dangerous to the same—then such structures have the same relation to that agency which machines have to man ; and in this aspect it may be a legitimate figure of speech to call them living machines. What these machines do is different in kind from the things which human machines do, but both are alike in this—that what- ever they do is done in virtue of their construction and of the powers which have been given to them by the mind which made them.

Lastly, the reason of man himself is an actual illustration of mind and will as an efficient power in nature, and implies a creative mind. We cannot imagine the development of reason from that which has no reason, and must admit that only the 'inspiration of the Almighty' could have given understanding. The inherent absurdity of the evolution of powers and properties from things in which they are not even potentially contained appears nowhere more clearly than here. The subject is, however, sufficiently im- portant to demand a separate chapter.

MAN IN NATURE

FEW words are used among us more loosely than 'nature.' Sometimes it stands for the material universe as a whole. Sometimes it is personified as a sort of goddess, working her own sweet will with material things. Sometimes it expresses the forces which act on matter, and again it stands for material things themselves. It is spoken of as subject to law, but just as often natural law is referred to in terms which imply that nature itself is the lawgiver. It is supposed to be opposed to the equally vague term 'supernatural'; but this term is used not merely to denote things above and beyond nature, if there are such, but certain opinions held respecting natural things. On the other hand, the natural is contrasted with the artificial, though this is always the outcome of natural powers and is certainly not supernatural. Again, it is applied to the inherent properties of beings for which we are unable to account, and which we are content to say constitute their nature. We cannot look into the works of any of the more speculative

[1] The substance of this chapter was first published in the *Princeton Review*.

writers of the day without meeting with all these uses of the word, and have to be constantly on our guard, lest by a change of its meaning we shall be led to assent to some proposition altogether unfounded.

For illustrations of this convenient though dangerous ambiguity, I may turn at random to almost any page in Darwin's *Origin of Species.* In the beginning of Chapter III. he speaks of animals ' in a state of nature,' that is, not in a domesticated or artificial condition, so that here nature is opposed to the devices of man. Then he speaks of species as ' arising in nature,' that is, spontaneously produced in the midst of certain external conditions or environment outside of the organic world. A little farther on he speaks of useful varieties as given to man by ' the hand of Nature,' which here becomes an imaginary person ; and it is worthy of notice that in this place the printer or proof-reader has given the word an initial capital, as if a proper name. In the next section he speaks of the ' works of nature ' as superior to those of art. Here the word is not only opposed to the artificial, but seems to imply some power above material things and comparable with or excelling the contriving intelligence of man. I do not mean by these examples to imply that Darwin is in this respect more inaccurate than other writers. On the contrary, he is greatly surpassed by many of his contemporaries in the varied and fantastic uses of this versatile word. An illustration which occurs to me

here, as at once amusing and instructive, is an expression used by Romanes, and which appears to him
to give a satisfactory explanation of the mystery of
elevation in nature. He says, 'Nature selects the
best individuals out of each generation to live.' Here
nature must be an intelligent agent, or the statement
is simply nonsensical. The same alternative applies
to much of the use of the favourite term 'natural
selection.' In short, those who use such modes of
expression would be more consistent if they were at
once to come back to the definition of Seneca, that
nature is ' a certain divine purpose manifested in the
world.'

The derivation of the word gives us the idea of
something produced or becoming, and it is curious
that the Greek *physis*, though etymologically distinct,
conveys the same meaning—a coincidence which may
perhaps lead us to a safe and serviceable definition.
Nature rightly understood is, in short, an orderly
system of things in time and space, and this not invariable, but in a state of constant movement and
progress, whereby it is always becoming something
different from what it was. Now man is placed in
the midst of this orderly, law-regulated yet ever-pro-
gressive system, and is himself a part of it ; and if
we can understand his real relations to its other parts,
we shall have made some approximation to a true
philosophy. If, with Tyndall, we were to place man
outside of nature, then the human mind would at once

become to us a supernatural intelligence. But truth forbids such a conclusion. The reason of man, though far beyond the intelligence of other animals, so harmonises with natural laws, and acts in such uniformity with these, that it is evidently a part of the great unity of nature, and we cannot, without violence, dissociate man from nature. If we could do so, we should have good ground to distrust all the conclusions of our own reason, in so far as they relate to the material universe. In short, we should cut away the foundations of science, and what remained of religion would be preternatural, in the bad sense of destroying the unity of nature, and with it our confidence in the unity of God.

It may be well to remark here that this consideration limits and defines our use of the much-abused word ' supernatural,' which perhaps it would be well for us to follow the example of our Christian Scriptures in avoiding altogether as a misleading term. If by supernatural we mean something outside of and above nature and natural law, there is really no such thing in the universe. There is no doubt that which is, ' spiritual,' as distinguished from that which is natural in the material sense, but the spiritual has its own laws, which are not in conflict with those of the natural. Even God cannot in this sense be said to be supernatural, since His will is in strict conformity with natural law. Yet this absurd sense of the term supernatural is constantly employed both by the enemies and friends of religion, to the disgust of all clear thinkers. The only true sense in which any being or thing can be said to be supernatural is that in which we use it with reference to the creation of matter or energy or the constitution of natural law. The power which caused these things is above nature, but not outside of nature, for matter, energy, and law must be included in, and in harmony with, the creative will.

To return from this digression, if man is a part of nature, then we see how not only his bodily organism conforms to natural structures and laws, but how his mind is in harmony with the external world, so that he can comprehend it, enter into it, and utilise it for his own purposes. Even his moral and religious ideas must in this case be more or less adapted to his conditions of existence as a part of nature. We have here also a sure guarantee for the correctness of our perceptions and of our conclusions respecting the laws of nature. In like manner, there is here a sense in which man is above nature, because he is placed at the head of it. In another sense he is inferior to the aggregate of nature, because, as Agassiz well puts it, there is in the universe a 'wealth of endowment of the most comprehensive mental manifestations' which he can never fully comprehend.

Still further, if the universe has been created, then just as its laws must be in harmony with the will of the Creator, so must our mental constitution ; and man as a reasoning and conscious being must be made in the image of his Maker. If we discard the idea of an intelligent Creator, then mind and all its powers must be potentially in the atoms of matter or in the forces which move them ; but this is a mere form of words, and most unscientific, since it requires us to attribute to matter properties which experiment does not show it to possess. Thus the existence of man is not only a positive proof of mind in nature, but affords the strongest possible evidence of a higher creative mind, from which that of man emanates. Even on the principle of evolution, no lower power could have produced the universe than the mind which has been evolved from it, and the power which did this must have been at least as much greater, and more intelligent, as the universe exceeds human power and human capacities to fathom its mysteries. Thus we return to the Pauline idea that the power and divinity of the Creator are proved by the works which He has made. Legitimate science can say nothing more and nothing less.

But even Science may be permitted to point to what lies beyond her domain, and to indicate the probability that the God who has in the long geologic ages fitted the earth for man, and endowed it with so many evidences of His own power and wisdom, and who has made us in His own image, has not left us as orphans, but has given us a revelation of His will, and has provided for us a Saviour from all the sins and evils that afflict humanity.

Regarding man, then, as a part of nature, we must hold to his entering into the grand unity of the natural system, and must not set up imaginary antagonisms between man and nature, as if he were outside of it. An instance of this appears in Tyndall's celebrated Belfast address, where he says, in explanation of the errors of certain of the older philosophers, that 'the experiences which formed the weft and woof of their theories were chosen not from the study of nature, but from that which lay much nearer to them—the observation of Man'; a statement this which would make man a supernatural or at least a preternatural being. Again, it does not follow that because man is a part of nature that he must be precisely on a level with its other parts. There are in nature many planes of existence, and man is no doubt on one of its higher planes and possesses distinguishing powers and properties of his own. Nature, like a perfect organism, is not all eye or all hand, but includes various organs, and, so far as we see it in our planet, man is its head, though we can easily conceive that there may be higher beings in other parts of the universe beyond our ken.

The view which we may take of man's position relatively to the beings which are nearest to him, namely, the lower animals, will depend on our point of sight—whether that of mere anatomy and physio-

logy, or that of psychology and pneumatology as
well. This distinction is the more important, since
under the somewhat delusive term ' biology ' it has
been customary to mix up all these considerations ;
while on the other hand those anatomists who regard
all the functions of organic beings as mechanical
and physical, do not scruple to employ this term
biology for their science, though on their hypothesis
there can be no such thing as life, and consequently
the use of the word by them must be either superstitious or hypocritical.

Anatomically considered, man is an animal of the
class *Mammalia.* In that class, notwithstanding the
heroic efforts of some modern detractors from his
dignity to place him with the monkeys in the order
Primates, he undoubtedly belongs to a distinct order.
I have elsewhere argued that if he were an extinct
animal, the study of the bones of his hand or of his
head would suffice to convince any competent palæon-
tologist that he represents a distinct order, as far
apart from the highest apes as they are from the car-
nivora. That he belongs to a distinct family no
anatomist denies, and the same unanimity of course
obtains as to his generic and specific distinctness.
On the other hand, no zoological systematist now
doubts that all the races of men are specifically identical. Thus we have the anatomical position of man
firmly fixed in the system of nature, and he must be
content to acknowledge his kinship not only with the

O

higher animals nearest to him, but with the humblest animalcule. With all he shares a common material, and many common features of structure.

When we ascend to the somewhat higher plane of physiology we find in a general way the same relationship to animals. Of the four grand leading functions of the animal—nutrition, reproduction, voluntary motion, and sensation—all are performed by man as by other animals. Here, however, there are some marked divergences connected with special anatomical structures on the one hand and with his higher endowments on the other. With regard to food, for example, man might be supposed to be limited by his masticatory and digestive apparatus to succulent vegetable substances. But by virtue of his inventive faculties he is practically unlimited, being able by artificial processes to adapt the whole range of vegetable and animal food substances to his use. He is very poorly furnished with natural tools to aid in procuring food, as claws, tusk, &c., but by invented implements he can practically surpass all other creatures. The long time of helplessness in infancy, while it is necessary for the development of his powers, is a practical disadvantage which leads to many social arrangements and contrivances specially characteristic of man. Man's sensory powers, while inferior in range to those of many other animals, are remarkable for balance and completeness, leading to perceptions of differences in colours, sounds, &c. which lie at the foundation of art. The specialisation of the hand again connects itself with contrivances which render an animal naturally defenceless the most formidable of all, and an animal naturally gifted with indifferent locomotive powers able to outstrip all others in speed and range of locomotion. Thus the physiological endowments of man, while common to him with other animals, and in some respects inferior to theirs, present, in combination with his higher powers, points of difference which lead to the most special and unexpected results.

In his psychical relations, using this term in its narrower sense, we may see still greater divergences from the line of the lower animals. These may no doubt be connected with his greater volume of brain ; but recent researches seem to show that brain has more to do with motor and sensory powers than with those that are intellectual, and thus that a larger brain is only indirectly connected with higher mental manifestations. Even in the lower animals it is clear that the ferocity of the tiger, the constructive instinct of the beaver, and the sagacity of the elephant depend on psychical powers which are beyond the reach of the anatomist's knife ; and this is still more markedly the case in man. Following in part the ingenious analysis of Mivart, we may regard the psychical powers of man as reflex, instinctive, emotional, and intellectual ; and in each of these aspects we shall find points of resemblance to other animals and of divergence from them. In regard to reflex actions, or those which are merely automatic, inasmuch as they are intended to provide for certain important functions without thought or volition, their develop- ment is naturally in the inverse ratio of psychical ele- vation, and man is consequently in this respect in no way superior to lower animals.

The same may be said with reference to instinc- tive powers, which provide often for complex actions in a spontaneous and unreasoning manner. In these also man is rather deficient than otherwise ; and since from their nature they limit their possessors to narrow ranges of activity, and fix them within a definite scope of experience and efficiency, they would be incompatible with those higher and more versatile inventive powers which man possesses. The comb- building instinct of the bee, the nest-weaving instinct of the bird, are fixed and invariable things, obviously incompatible with the varied contrivance of man ; and while instinct is perfect within its narrow range, it cannot rise beyond this into the sphere of unlimited thought and contrivance. Higher than mere instinct are the powers of imagination, memory, and associa- tion, and here man at once steps beyond his animal associates, and develops these in such a variety of ways that even the rudest tribes of men, who often appear to trust more to these endowments than to higher powers, rise into a plane immeasurably above that of the highest and most intelligent brutes, and toward which they are unable, except to a very limited degree, to raise those of the more domesticable animals which they endeavour to train into companionship with themselves. It is, however, in these domesticated animals that we find the highest degree of approximation to ourselves in emotional development, and this is perhaps one of the points that fits them for such human association. In approaching the higher psychical endowments, the affinity of man and the brute appears to diminish and at length to cease, and it is left to him alone to rise into the domain of the rational and ethical.

Those supreme endowments of man we may, following the nomenclature of ancient philosophy and of our sacred Scriptures, call 'pneumatical' or spiritual. They consist of consciousness, reason, and moral volition. That man possesses these powers everyone knows ; that they exist or can be developed in lower animals no one has succeeded in proving. Here at length we have a severance between man and material nature. Yet it does not divorce him from the unity of nature, except on the principles of atheism. For if it separates him from animals it allies him with the Power who made and planned the animals. To the naturalist the fact that such capacities exist in a being who in his anatomical structure so closely resembles the lower animals, constitutes an evidence of the independent existence of those powers, and of their spiritual character and rela-

tion to a higher power, which, I think, no metaphysical reasoning or materialistic scepticism will suffice to
invalidate. It would be presumption, however, from
the standpoint of the naturalist to discuss at length
the powers of man's spiritual being. I may refer
merely to a few points which illustrate at once his
connection with other creatures, and his superiority to
them as a higher member of nature.

And first we may notice those axiomatic beliefs
which lie at the foundation of human reasoning, and
which, while apparently in harmony with nature, do not
admit of verification except by an experience impossible to finite beings. Whether these are ultimate
truths, or merely results of the constitution bestowed
on us, or effects of the direct action of the creative mind
on ours, they are to us like the instincts of animals—
infallible and unchanging. Yet just as the instincts of
animals unfailingly connect them with their surroundings, our intuitive beliefs fit us for understanding
nature and for existing in it as our environment.
These beliefs also serve to connect man with his fellow man ; and in this aspect we may associate with
them those universal ideas of right and wrong, of
immortality, and of powers above ourselves, which
pervade humanity.

Another phase of this spiritual constitution is
illustrated by the ways in which man, starting from
powers and contrivances common to him and animals,
develops them into new and higher uses and results.

This is markedly seen in the gift of speech. Man,
like other animals, has certain natural utterances ex-
pressive of emotions or feelings. He can also, like
some of them, imitate the sounds produced by ani-
mate or inanimate objects, and he has better mechanical powers of articulation than other animals. But
when he develops these gifts into a system of speech,
expressing not mere sounds occurring in nature, but,
by association and analogy with these, properties and
relations of objects, and general and abstract ideas, he
rises into the higher sphere of the spiritual. He thus
elevates a power of utterance common to him with
animals to a higher plane, and, connecting it with his
capacity for understanding nature and arriving at
general truths, asserts his kinship to the great creative
mind, and furnishes a link of connection between the
material universe and the spiritual Creator.

The mode of existence of man in nature is as
well illustrated by his arts and inventions as by anything else ; and these serve also to enlighten us as to
the distinction between the natural and the artificial.
Naturalists often represent man as dependent on
nature for the first hints of his useful arts. There
are in animal nature tailors, weavers, masons, potters,
carpenters, miners, and sailors, independently of man,
and many of the tools, implements, and machines which
he is said to have invented were perfected in the structures of lower animals long before he came into existence. In all these things man has been an assiduous

learner from nature, though in some of them, as for
example in the art of aerial navigation, he has striven
in vain to imitate the powers possessed by other animals. But it may well be doubted whether man is in
this respect so much an imitator as has been supposed,
and whether the resemblance of his plans to those
previously realised in nature does not depend on that
general fitness of things which suggests to rational
minds similar means to secure similar ends. But in saying this we in effect say that man is not only a part of
nature, but that his mind is in harmony with the plans
of nature, or, in other words, with the methods of
the creative mind. Man is also curiously in harmony
with external nature in the combination in his works
of the ideas of plan and adaptation, of ornament and
use. In architecture, for example, devising certain
styles or orders, and these for the most part based on
imitations of natural things, he adapts these to his
ends, just as in nature types of structure are adapted
to a great variety of uses ; and he strives to combine,
as in nature, perfect adaptation to use with conformity
to type or style. So in his attempts at ornament he
copies natural forms, and uses these forms to decorate
or conceal parts intended to serve essential purposes
in the structure. This is at least the case in the
purer styles of construction. It is in the more debased
styles that arches, columns, triglyphs, or buttresses
are placed where they can serve no useful purpose,
and become mere excrescences. But in this case the

abnormality resulting breeds in the beholder an un-
pleasing mental confusion, and causes him, even when
he is unable to trace his feelings to their source, to be
dissatisfied with the result. Thus man is in harmony
with that arrangement of nature which causes every
ornamental part to serve some use, and which unites
adaptation with plan.

The following of nature must also form the basis
of those fine arts which are not necessarily connected
with any utility ; and in man's pursuit of art of this
kind we see one of the most recondite and at first
sight inexplicable of his correspondences with the
other parts of nature, for there is no other creature
that pursues art for its own sake. Modern archæo-
logical discovery has shown that the art of sculpture
began with the oldest known races of man, and that
they succeeded in producing very accurate imitations
of natural objects. But from this primitive starting-
point two ways diverge. One leads to the conven-
tional and the grotesque, and this course has been
followed by many semi-civilised nations. Another
leads to accurate imitation of nature, along with new
combinations arising from the play of intellect and
imagination. Let us look for a moment at the actual
result of the development of these diverse styles of art,
and at their effect on the culture of humanity as existing in nature. We may imagine a people who have
wholly discarded nature in their art, and have devoted
themselves to the monstrous and the grotesque. Such a

people, so far as art is concerned, separates itself widely from nature and from the mind of the Creator, and its taste and possibly its morals sink to the level of the monsters it produces. Again, we may imagine a people in all respects following nature in a literal and servile manner. Such a people would probably attain to but a very moderate amount of culture but having a good foundation, it might ultimately build up higher things. Lastly, we may fancy a people who, like the old Greeks, strove to add to the copying of nature a higher and ideal beauty, by combining in one the best features of many natural objects, or devising new combinations not found in nature itself. In the first of these conditions of art we have a falling away from or caricaturing of the beauty of nature. In the second we have merely a pupilage to nature. In the third we find man aiming to be himself a creator, but basing his creations on what nature has given him. Thus all art worthy of the name is really a development of nature. It is true the eccentricities of art and fashion are so erratic that they may often seem to have no law. Yet they are all under the rule of nature ; and hence even uninstructed common-sense, unless dulled by long familiarity, detects in some degree their incongruity, and though it may be amused for a time, at length becomes wearied with the mental irritation and nervous disquiet which they produce.

I may be permitted to add that all this applies with still greater force to systems of science and philosophy. Ultimately these must all be tested by the verities of nature to which man necessarily submits his intellect, and he who builds for aye must build on the solid ground of nature. The natural environment presents itself in this connection as an educator of man. From the moment when infancy begins to exercise its senses on the objects around, this education begins—training the powers of observation and comparison, cultivating the conception of the grand and beautiful, leading to analysis and abstract and general ideas. Left to itself, it is true this natural education extends but a little way, and ordinarily it becomes obscured or crushed by the demands of a hard utility, or by an artificial literary culture, or by the habitude of monstrosity and unfitness in art. Yet when rightly directed it is capable of becoming an instrument of the highest culture, intellectual, æsthetic, and even moral. I have in writing on evolution in education insisted on the importance of following nature in the education of the young, and of dropping much that is arbitrary and artificial. Here I would merely remark that when we find that the accurate and systematic study of nature trains most effectually some of the more practical powers of mind, and leads to the highest development of taste for beauty in art, we see in this relation the unity of man and nature, and the unity of both with something higher than either.

It may, however occur to us here that when we

consider man as an improver and innovator in the
world, there is much that suggests a contrariety between him and nature, and that instead of being the
pupil of his environment he becomes its tyrant. In
this aspect man, and especially civilised man, appears
as the enemy of wild nature, so that in those districts
which he has most fully subdued many animals and
plants have been exterminated, and nearly the whole
surface has come under his processes of culture, and
has lost the characteristics which belonged to it in its
primitive state. Nay, more, we find that by certain
kinds of so-called culture man tends to exhaust and
impoverish the soil, so that it ceases to minister to
his comfortable support and becomes a desert. Vast
regions of the earth are in this impoverished condition, and the westward march of exhaustion warns
us that the time may come when even in comparatively new countries like America the land will cease
to be able to sustain its inhabitants. Behind this
stands a still farther and portentous possibility. The
resources of chemistry are now being taxed to the
utmost to discover methods by which the materials of
human food may be produced synthetically ; and we
may possibly at some future time find that albumen
and starch may be manufactured cheaply from their
elements by artificial processes. Such a discovery
might render man independent of the animal and
vegetable kingdoms. Agriculture might become an
unnecessary and unprofitable art. A time might

come when it would no longer be possible to find
a green field, a forest, or a wild animal ; and when
the whole earth would be one great factory, in which
toiling millions were producing all the materials
of food, clothing, and shelter. Such a world may
never exist, but its possible existence may be
imagined, and its contemplation brings vividly
before us the vast powers inherent in man as a
subverter of the ordinary course of nature. Yet even
this ultimate annulling of wild nature would be
brought about, not by anything preternatural in man,
but simply by his placing himself in alliance with
certain natural powers and agencies, and by their
means attaining dominion over the rest.

Here there rises before us a spectre which science
and philosophy appear afraid to face, and which asks
the dread question, What is the cause of the apparent
abnormality in the relations of man and nature ? In
attempting to solve this question we must admit
that the position of man even here is not without
natural analogies. The stronger preys upon the
weaker, the lower form gives place to the higher, and
in the progress of geological time old species have
died out in favour of newer, and old forms of life have
been exterminated by later successors. Man, as the
newest and highest of all, has thus the natural right
to subdue and rule the world. Yet there can be little
doubt that he uses this right unwisely and cruelly,
and these terms themselves explain why he does so,

because they imply freedom of will. Given a system of nature destitute of any being higher than the instinctive animal, and introduce into it a free rational agent, and you have at once an element of instability. So long as his free thought and purpose continue in unison with the arrangements of his environment, so long all will be harmonious; but the very hypothesis of freedom implies that he can act otherwise; and so perfect is the equilibrium of existing things that one wrong or unwise action may unsettle the nice balance, and set in operation trains of causes and effects producing continued and ever-increasing disturbance. This 'fall of man,' we know as a matter of observation and experience, has actually occurred; and it can be retrieved only by casting man back again into the circle of merely instinctive action, or by carrying him forward until by growth in wisdom and knowledge he becomes fitted to be the lord of creation. The first method has been proved unsuccessful by the rebound of humanity against all the attempts to curb and suppress its liberty. The second has been the effort of all reformers and philanthropists since the world began; and its imperfect success affords a strong ground for clinging to the theistic view of nature, for soliciting the intervention of a Power higher than man, and for hoping for a final restitution of all things through the intervention of that Power. Mere materialistic evolution must ever and necessarily fail to account for the higher nature of man, and also for his moral aberrations. These only come rationally into the system of nature under the supposition of a Higher Intelligence, from whom man emanates and whose nature he shares.

But on this theistic view we are introduced to a kind of unity and of evolution for a future age, which is the great topic of revelation, and is not unknown to science and philosophy, in connection with the law of progress and development deducible from the geological history, in which an ascending series of lower animals culminates in man himself. Why should there not be a new and higher plane of existence to be attained to by humanity—a new geological period, so to speak, in which present anomalies shall be corrected, and the grand unity of the universe and its harmony with its Maker fully restored? This is what Paul anticipates when he tells us of a 'pneumatical' or spiritual body to succeed to the present natural or 'psychical' one, or what Jesus Himself tells us when He says that in the future state we shall be like to the angels. Angels are not known to us as objects of scientific observation, but such an order of beings is quite conceivable, and this not as super-natural, but as part of the order of nature. They are created beings like ourselves, subject to the laws of the universe, yet free and intelligent and liable to error, in bodily constitution freed from many of the limitations imposed on us, mentally having higher range and grasp, and consequently masters of natural powers not under our control. In short, we have here pictured to us an order of beings forming a part of nature, yet in their powers as miraculous to us as we might be supposed to be to lower animals, could they think of such things.

This idea of angels bridges over the otherwise impassable gulf between humanity and deity, and illustrates a higher plane than that of man in his present state, but attainable in the future. Dim perceptions of this would seem to constitute the substratum of the ideas of the so-called polytheistic religions. Christianity itself is in this aspect not so much a revelation of the supernatural as the highest bond of the great unity of nature. It reveals to us the perfect man who is also one with God, and the mission of this divine man to restore the harmonies of God and humanity, and consequently also of man with his natural environment in this world, and with his spiritual environment in the higher world of the future. If it is true that nature now groans because of man's depravity, and that man himself shares in the evils of this dis-harmony with nature around him, it is clear that if man could be restored to his true place in nature he would be restored to happiness and to harmony with God ; and if, on the other hand, he can be restored to harmony with God, he will then be restored also to harmony with his natural environment, and so to life and happiness and immortality. It is here that the old story of Eden, and the teaching of Christ, and the prophecy of the New Jerusalem strike the same note which all material nature gives forth when we interrogate it respecting its relations to man. The profound manner in which these truths appear in the teaching of Christ has perhaps not been appreciated as it should, because we have not sought in that teaching the philosophy of nature which it contains. When He points to the common weeds of the fields, and asks us to consider the garments more gorgeous than those of kings in which God has clothed them, and when He says of these same wild flowers, so daintily made by the Supreme Artificer, that to-day they are, and to-morrow are cast into the oven, He gives us not merely a lesson of faith, but a deep insight into that want of unison which, centring in humanity, reaches all the way from the wild flower to the God who made it, and requires for its rectification nothing less than the breathing of that Divine Spirit which first evoked order and life out of primeval chaos. When He points out to us the growth of these flowers without any labour of their own, He opens up one of the most profound analogies between the growth of the humblest living thing and that of the new spiritual nature which may be planted in man by that same Divine Spirit.

CHAPTER X

WE have already seen that, agnostics themselves being judges, man must have a religion, and that if he makes the material universe the highest object of veneration this must be to him his God, while if he is content to take humanity as his highest ideal, he must look for the best possible manifestations of human nature, else his religion can have no elevating power. To the theist the universe is not in itself God, but may testify to God as its Creator; to the Christian the noblest ideal of humanity along with divinity is the Lord Jesus Christ.

It is evident, however, that the current Darwinian and Neo-Lamarckian forms of evolution fall entirely short of what even the agnostic may desiderate as religion.

If the universe is causeless and a product of fortuitous variation and selection, and if there is no design or final cause apparent in it, it becomes literally the enthronement of unreason, and can have no claims to the veneration or regard of an intelligent being. If man is merely an accidentally improved descendant of apes, his intuitions and decisions as to things unseen must be valueless and unfounded. Hence it is a lamentable fact that the greater part of evolutionist men of science openly discard all religious belief, and teach this unbelief to the multitude who cannot understand the processes by which it is arrived at, but who readily appreciate the immoral results to which it leads in the struggle for existence or the stretching after material advantages.

It is true that there may be a theistic form of evolution, but let it be observed that this is essentially distinct from Darwinism or Neo-Lamarckianism. It postulates a Creator, and regards the development of the universe as the development of His plans by secondary causes of His own institution. It necessarily admits design and final cause. It can even set up plausible analogies between the supposed material development and that which is moral and spiritual, many of which are, however, based on misstatements as to natural facts. The weakness of this position consists in the objections to the doctrine of evolution itself as a means of explaining nature, and in the incongruity between the methods supposed by evolution and the principles of design, finality, and ethical purity inseparable from a true and elevating religion. The theistic evolutionists have also before them the danger that in the constant flux of philosophic opinion they will find their system of theology, which at present rides so triumphantly on the flood-tide of a popular movement, eventually stranded, as so many older ones have been, on the sandbanks of the ebb.

It will therefore be the safest as well as the most candid and truthful course, both for the scientific worker and the theologian, to avoid committing himself to any of the current forms of evolution. The amount of assumption and reasoning in a vicious circle involved in these renders it certain that none of them can long survive. On the other hand, the extensive investigations as to facts, and the varied discussions which have arisen out of Darwinism, cannot fail to leave an impress on science and to increase our knowledge, at least as to the modes of creative development. The winnowing process has already begun, and our immediate successors may be able to secure the pure grains of truth after the chaff of unproved hypotheses has been swept away.

Looking to this desirable result, there are certain principles that arise out of the previous discussion to which we may firmly hold without fear of being dislodged by any assailant.

1. No system of the universe can dispense with a First Cause, eternal and self-existent ; and the First Cause must necessarily be the living God, whose will is the ultimate force and the origin of natural law. Our knowledge of God cannot be direct, but must be mediate, either through His works as Creator or

through such revelation as He may have made of Himself to the human mind.

2. In studying natural things we must keep before our minds the certainty that the laws which we can ascertain have no validity except as expressions of the power behind nature. Consequently the reference of any effect to a secondary cause or the ascertaining of the law of operation of such cause in no respect diminishes the dependence of the whole on the Divine will.

3. While we are justified in taking an anthropomorphic view of the operations of God as being ourselves spiritually in His image, we must bear in mind that in many important respects He must infinitely transcend us and our modes of thought. To Him time and space are not limitations as to us, and the microscopically small may be relatively as great as that which seems to us almost infinitely large. It is sometimes represented as derogatory to God that He should paint the petals of flowers; but with Him this is not painting. He deals with things invisible to the human painter, with the individual cells, with the pigment which they contain, with the arrangement of the atoms and molecules that make the pigment. To Him the arrangement of a multitude of atoms to make a microscopic dot of pigment must be neither a greater nor less work than the ordering of a system of worlds. The immensity of the universe can in no respect distract His attention from the humblest atom, because He is present and efficient in all.

4. We have already seen that it results from this that material nature cannot fully reveal God to us. Our present knowledge of nature is, as we too well know, relatively very small. But even if we could know, and have distinctly before our minds every fact and law of the whole universe and all their relations and interactions, we should on the whole have only one set of possibilities out of an infinite number ; and, as we have already seen, the manifestation of God would be in a manner which must be in many respects the converse of His essential properties. A photograph represents to me a friend, but it is not the friend himself ; a building represents an architect, but it is not the architect. It would seem as if in many current arguments respecting agnosticism these simple principles were altogether overlooked.

5. Creation was not an instantaneous process, but extended through periods of vast duration. In every stage we may rest assured that God, like a wise builder, used every previous course as a support for the next ; that He built each succeeding storey of the wonderful edifice on that previously prepared for it ; and that His plan developed itself as His work proceeded. So far, there must have been evolution and development. But the attempt to narrow this plan to any one little principle that we have laboriously worked out must be futile. Such analogies, even if well founded in nature, can only be partial and limited in application, and nothing can be really

gained by an enthusiastic application of them beyond their legitimate bounds. The present condition of the Darwinian doctrine of natural selection clearly proves this, and the various substitutes for it, or additions to it, now proposed are all equally partial. Instead of regarding any of these theories as final or sufficient, we should scrutinise them as to their validity and extent of application, and we shall find that, in so far as any of them have reality, they cover only a few facts, and still leave a boundless region to be explored, even with reference to the modes of the development.

6. Even our ideas of design and final cause must be held in subjection to the infinite nature of God. Crude views on these subjects have, perhaps, aided in producing present scepticism as to natural theology. When Hegel says that all nature is final cause, and that it is not necessary to conceive of final cause as it exists in our consciousness, he does not necessarily imply that nature itself is God, but that God's design as manifested in nature is only in a small part intelligible to us. We are constantly discovering new uses and adaptations previously unknown ; and in the Divine mind there must be infinite designs and objects as yet quite inaccessible to us. We may learn this by a moment's thought of the development in geological time. An intelligent observer introduced to the earth when tenanted only by aquatic invertebrates would reason as to this as a finality ; but he

might by no means be able to divine the plan and design of the Creator to be afterwards realised in the vertebrate animals and in man. Thus both in amount and in time the design and final cause in nature would be very partially conceivable by him. This is what the Bible means when it points to the glory of God Himself as the final cause ; and we can well imagine that this glory may shine with infinitely greater effulgence before the minds of higher intelligences, or before our own minds in the future.

The same consideration helps us to understand how we may be disposed even to condemn as imperfect the arrangements of nature. The time was when ferocious wild beasts were the lords of the earth ; but they and their doings are to be judged by the laws of their own order, not by the higher ideal of our sentiments or of the nature of God, and their uses are to be gauged by the future that was to succeed them as well as by their own time. Man himself has failed so far to realise the highest ideal embodied in his nature and capacities ; but God has farther designs with humanity, to be realised in the 'manifestation of the sons of God.' We may thus learn that, while God gives life, there is a struggle, not for existence, but against evil ; and we may have faith that the fit will survive in the highest and best sense of an approving conscience now and in the coming glory hereafter. Thus far our own conscience and natural religion may carry us.

7. It follows that the material universe, while, by the power and divine attributes present in it, our minds may be enlarged and elevated, cannot fully satisfy the demands of our religious life. However we may be instructed and elevated by the marvellous exhibition of divinity in nature ; however God may shine within ourselves by the light of conscience, we must find ourselves surrounded by those inscrutable mysteries with which the great minds of antiquity so manfully strove, and which are so clearly presented to us in the discussions of Job with his friends. Rightly regarded, even these mysteries may, by analogy with God's natural procedure, be to some extent solved, as they were by the patriarch of Uz, when, in contemplating the marvellous works of nature, he humbled himself before God and repented in dust and ashes. It is plain that if it has pleased God to reveal Himself directly to man, in addition to the indirect revelation of nature, and to the testimony of our own moral intuitions, this must be a great gain. Hence men have yearned for such revelation, and have believed that it has been given by the Spirit of God in the visions of prophets and the narratives of holy men of old, and in these last times by the divine Son of God Himself. To Jesus Christ all men must turn, trusting to Him for salvation, and looking forward to the ultimate finality in His coming kingdom, in which only can be perfectly manifested the great designs of the Almighty Father

APPENDIX I

WEISMANN ON HEREDITY

THE new views advanced by Weismann have, while these pages were going through the press, been subjects of warm discussion in England, where his essays have been translated and republished ; but the subject has been so beclouded with technicalities and references to obscure facts of reproduction, that it is scarcely intelligible to non-technical readers. An explanation of the actual nature and bearing of these views may therefore be useful.

The subject may be regarded from the point of view either (1) of the facts of reproduction, or (2) of observed phenomena of inheritance.

1. With reference to the first of these, nothing is more certain than that in all animals, except a few of the lowest, there are special organs of reproduction, and that in these organs alone resides the power of permanent continuance of the species. The facts of budding and spontaneous division in some animals of low grade may be regarded as of only temporary importance. Farther, the organ of reproduction resolves itself into a single microscopic embryo cell or germ, a minute vesicle containing protoplasmic matter, fertilised by another or sperm-cell, and finally into the speck of proto-

plasm constituting the nucleus of this embryo cell. This minute living speck must contain in it potentially all the parts and organs that are produced from it. Weismann illustrates this in a clear manner by the observed fact of the spontaneous division of this nucleus into a vast number of separate granules, each of which plays a part in the formation of some portion of the embryo animal.

From this simple statement it follows beyond controve that any cause which effects a change in the structure properties of the future individual must pre-exist in t germ, and that any effect of external causes on the ad animal can have no effect in this respect unless it has modified the germ or germs of the new generation.

This conclusion, to which no physiologist can reasonably object, if it does not altogether subvert, as some think, the Darwinian and Lamarckian doctrines of evolution, at least weakens their force and diminishes their extent, and besides goes quite behind them into a region of antecedent causes, and shows that they must be of secondary value and importance. Weismann also insists very strongly on what he calls the 'immortality'—or properly, perpetuity — of germinal matter embodying the characters of the species—a very important and valuable idea.

All this leads us, however, not so much to deny that any causes acting on the adult can modify its progeny, as to inquire what causes, if any, can so profoundly affect the organism as to modify its germinal matter.

On this question we may first remark that such causes seem often to be psychical rather than material—that is, causes affecting the imagination and emotions. This has been known at least since the time of Jacob's experiment with Laban's cattle, and probably long before. It is still a matter of every-day experience both in man and animals, and opens a wide and inviting field of study, leading, perhaps, to more profound views of the causes of variation than any heretofore promulgated.

Again, since the germinal matter itself must be nourished by the common blood of the animal, it is possible that it may be influenced by any change by which this may be effected. Popular speech recognises this by speaking of certain tendencies as being in the blood of certain races of animals.

⁀ 2. Into the second field—that of study of actual phenomena—Weismann has largely entered as a matter of experiment and observation, and his conclusion is that, as a rule, characters impressed on the individual by accident or external influences are not perpetuated. Those that are congenital, being those that have originated in the germinal matter of the parent, have alone the power to affect the germinal matter of the offspring. If, for example, an infant from some cause, probably unknown to us, is born with six fingers instead of five, this peculiarity, whether useful to it or not, is likely to be perpetuated in some individuals at least of the next generation. But if the child loses one of its fingers by accident, its children are no more likely on that account to be born with only four fingers ; or if it is trained to use its fingers deftly in playing on an instrument or working at any mechanical art, its children will not on that account have more lissom fingers than others. This last statement, however, should perhaps be taken with the limitation that if the use of the fingers is of such a character as to act strongly on the mental or psychical nature of the adult, or on the general system, it may in that case so affect its germinal matter as to act on the offspring. This kind of inheritance, as the Duke of Argyll has pointed out, is very apparent in some domestic animals, as in dogs.

It is curious that these conclusions of Weismann equally affect Darwinism and Lamarckianism. They indeed bring both these doctrines together, as mere modifications of one superficial view, not reaching to the actual origin of varietal difference. If Weismann is right, we can no longer speak of an herbivorous quadruped as making efforts to reach food above its head, and so acquiring a tendency to elongation of neck which may be transmitted to its offspring, so that they may become giraffes. Nor can we be content merely to suppose an accidental elongation in one individual to give it such advantage in the struggle for existence as to cause it alone to survive in times of scarcity, and to propagate its kind. Such suppositions must be altogether gratuitous and trifling, and we must look for deeper causes capable of affecting the germinal matter, if we wish to establish the possibility of such changes. At present, as already hinted, the only causes of this kind certainly known in higher animals are those of a psychical character, and this is perhaps one reason of the liability of the more intelligent and shifty animals to varietal change. The similar capacity of some animals low in the scale may depend merely on the wider scope of vital work in less differentiated organisms, or on the greater liability of the whole organism to be affected by any change.

Whatever may be the ultimate amount of acceptance of these remarkable and ingenious views of the German physio logist, they no doubt open a vista which extends far beyond the crude ideas of evolution at present current.

It is farther to be observed in this connection that the discussions above referred to relate to variation rather than to origin of species, that they do not establish any hard-and-fast line separating congenital from acquired characters, and that they strongly emphasise the objections against mere accident as a cause of variation, and show the necessity, in order to the origin and perpetuation of varieties, not merely of one change, but of many correlated changes. They also show that the changes supposed must take place by anticipation in the germinal matter before their utility or inutility can be proved, and they emphasise the obstacles set up by sexual reproduction against unlimited divergence from the specific type. All these points are being developed in the discussions now in progress, and they must, ere long, profoundly modify the views of biologists as to the existing theories of evolution.

APPENDIX II

DR. McCOSH ON EVOLUTION

THE venerable ex-president of Princeton has just issued (1890) a second edition of his little work, *The Development Hypothesis under a new name : The Religious Aspect of Evolution*. The work makes no serious attempt to prove the validity of any of those various and often conflicting theories of evolution, the insufficiency of which, regarded in the light of scientific causation, I have endeavoured to show in the preceding pages. It assumes them all as established scientific results, and then proceeds to show that they can be received up to a certain point without destroying our belief in God. Perhaps it would be correct to say that the actual thesis of the work is that the belief in secondary causes in creation is perfectly consistent with a belief in a Divine First Cause. This is very clearly stated, and with much interesting illustration ; and as setting forth this great principle the work is of value, and its use in this respect will remain, even if all those imaginary and partial causes of development on which it relies should be swept away as of no scientific validity, and replaced by more rational views of the vastly complicated and still mysterious causes which have no doubt conspired under Creative guidance to bring about the succession of living beings in geological time.   In this respect the work is similar in its tendency to Drummond's *Natural Law in the Spiritual World*; and in another aspect both may be regarded as examples of the tendency of theology to conform itself to the philosophical and scientific hypotheses which are ever cropping up and disappearing.   For a time such conformity carries all before it, but it incurs the danger that when the false or partial hypotheses have been discarded the high truths imprudently connected with them may be disc also.

I am reminded here, however, to express one obligation which the world owes, not so much to a ing system of evolution, as to the discussion and c those systems.   It is that attention has been direc manner never before witnessed, to the power of herec environment, of use and disuse in improving or deterior. humanity.   The bearing of this on the physical, me. and moral education and advancement of man is of i practical importance, and merits a more full discussion th it has yet received on the part of those who are not ev lutionists in the ordinary sense of the term, but who believe in development and in causation.

PRINTED BY
SPOTTISWOODE AND CO., NEW-STREET SQUARE
LONDON

www.ingramcontent.com/pod-product-compliance
Lightning Source LLC
Chambersburg PA
CBHW021949220326

41599CB00012BA/1430